YOUR KNOWLEDGE HAS VALUE

Bibliographic information published by the German National Library:

The German National Library lists this publication in the National Bibliography; detailed bibliographic data are available on the Internet at http://dnb.dnb.de .

Imprint:

Copyright © 2019 GRIN Verlag
Print and binding: Books on Demand GmbH, Norderstedt Germany
ISBN: 9783346085412

This book at GRIN:

https://www.grin.com/document/499425

Nagjibhai Rabari

Solved Problems on Differential Equations

GRIN Verlag

GRIN - Your knowledge has value

Since its foundation in 1998, GRIN has specialized in publishing academic texts by students, college teachers and other academics as e-book and printed book. The website www.grin.com is an ideal platform for presenting term papers, final papers, scientific essays, dissertations and specialist books.

Visit us on the internet:

http://www.grin.com/

http://www.facebook.com/grincom

http://www.twitter.com/grin_com

Q-1 Solve $y'' + xy' + y = 0$ near 0.

Given that $y'' + xy' + y = 0$ ....(1) Comparing (1) with $y'' + P(x)y' + Q(x)y = 0$,

Here $P(x) = x, Q(x) = 1$ Since $P(x) = x$ and $Q(x) = 1$ both are analytic at $x = 0$ is an ordinary point of (1)

To solve (1) , we take $y = c_0 + c_1 x + c_2 x^2 + c_3 x^3 + \ldots \ldots = \sum_{n=0}^{\infty} c_n x^n$ ....(2)

Differentiating (2) twice in succession with respect to 'x' , we get

$y' = \sum_{n=1}^{\infty} c_n n x^{n-1}$ and $y'' = \sum_{n=1}^{\infty} c_n n(n-1) x^{n-2}$ ....(3) Putting the

above values of y, y' and y'' in (1) $\sum_{n=1}^{\infty} c_n n(n-1) x^{n-2} + x \sum_{n=1}^{\infty} c_n n x^{n-1} + \sum_{n=0}^{\infty} c_n x^n = 0$ or

$\sum_{n=1}^{\infty} c_n n(n-1) x^{n-2} + \sum_{n=1}^{\infty} c_n n x^n + \sum_{n=0}^{\infty} c_n x^n = 0$ or

$\sum_{n=1}^{\infty} c_{n+2}(n+2)(n+1) x^n + \sum_{n=1}^{\infty} c_n n x^n + \sum_{n=0}^{\infty} c_n x^n = 0$ or

$[2c_2 + c_0] + \sum_{n=1}^{\infty} [c_{n+2}(n+2)(n+1) + n \, c_n + c_n] x^n = 0$ ....(4) Since (4) is an identity

, equating the constant term and the coefficient of various powers of x to zero , we get

$2c_2 + c_0 = 0 \Rightarrow c_2 = -\frac{c_0}{2}$ ....(5) and $c_{n+2}(n+2)(n+1) + n \, c_n + c_n = 0$; for all $n \geq 1$

....................(6) $\Rightarrow c_{n+2} = -\frac{c_n(n+1)}{(n+1)(n+2)} = -\frac{c_n}{(n+2)}$ ....(7) Putting $n = 1, 2, 3, \ldots\ldots$ in (7)

$c_3 = -\frac{c_1}{3}$, $c_4 = -\frac{c_2}{4} = \frac{c_0}{8}$, $c_5 = -\frac{c_3}{5} = \frac{c_1}{15}$, $c_6 = -\frac{c_4}{6} = -\frac{c_0}{48}$ .... Putting these values

in (1) , we get

$y = c_0(1 - \frac{x^2}{2} + \frac{x^4}{8} - \frac{x^6}{48} + \ldots\ldots\ldots) + c_1(x - \frac{x^3}{3} + \frac{x^5}{15} - \ldots\ldots\ldots)$; c_0 and c_1 are arbitrary constants.

Which is required solution.

Q-2 Solve $(1 - x^2)y'' - 2xy' + n(n+1)y = 0$ near 0 ; where n is constant.

Given that $(1 - x^2)y'' - 2xy' + n(n+1)y = 0$ ....(1) Dividing (1) by $(1 - x^2) \neq 0$ we

get, $y'' - \frac{2x}{(1-x^2)} y' + \frac{n(n+1)}{(1-x^2)} y = 0$ ....(2) Comparing (1) with $y'' + P(x)y' + Q(x)y = 0$,

Here $P(x) = -\frac{2x}{(1-x^2)}$, $Q(x) = \frac{n(n+1)}{(1-x^2)}$ Since $P(x) = -\frac{2x}{(1-x^2)}$, $Q(x) = \frac{n(n+1)}{(1-x^2)}$ both are analytic at $x = 0$ is an ordinary point of (1)

To solve (1) , we take $y = c_0 + c_1 x + c_2 x^2 + c_3 x^3 + \ldots\ldots = \sum_{m=0}^{\infty} c_m x^m$ ....(3)

Differentiating (3) twice in succession with respect to 'x' , we get

$y' = \sum_{m=1}^{\infty} c_m m x^{m-1}$ and $y'' = \sum_{m=1}^{\infty} c_m m(m-1) x^{m-2}$ ....(4)

Putting the above values of $y, y' and y''$ in (1)

$(1-x^2) \sum_{m=2}^{\infty} m(m-1) c_m x^{m-2}$ -2 $x \sum_{m=1}^{\infty} c_m m x^{m-1}$ + n (n+1) $\sum_{m=0}^{\infty} c_m x^m = 0$ or

$\sum_{m=2}^{\infty} m(m-1) c_m x^{m-2} - \sum_{m=2}^{\infty} m(m-1) c_m x^m$ -2 $\sum_{m=1}^{\infty} c_m m x^m$ + n (n+1) $\sum_{m=0}^{\infty} c_m x^m = 0$ or

$\sum_{m=0}^{\infty} (m+2)(m+1) c_{m+2} x^m - \sum_{m=2}^{\infty} m(m-1) c_m x^m$ -2 $\sum_{m=1}^{\infty} m c_m x^m$ + n (n+1) $\sum_{m=0}^{\infty} c_m x^m = 0$

or

$[2c_2 + n(n+1)c_0] + [6c_3 + n(n+1)c_1 - 2c_1]x + \sum_{m=2}^{\infty}[(m+2)(m+1)c_{m+2} + m(m-1) + 2m - n(n+1)c_m]x$

$= 0$

$[2c_2 + n(n+1)c_0] + [6c_3 + n(n+1)c_1 - 2c_1]x + \sum_{m=2}^{\infty}[(m+2)(m+1)c_{m+2} + (n^2 - m^2) + (n-m)c_m]x^m$

$= 0$

$[2c_2 + n(n+1)c_0] + [6c_3 + (n^2 + n - 2)c_1]x + \sum_{m=2}^{\infty}[(m+2)(m+1)c_{m+2} + (n^2 - m^2) + (n-m)c_m]x^m$

$= 0$

$[2c_2 + n(n+1)c_0] + [6c_3 + (n-1)(n+2)c_1]x + \sum_{m=2}^{\infty}[(m+2)(m+1)c_{m+2} + (n-m)(n+m+1)c_m]x^m = 0$

$[2c_2 + n(n+1)c_0] + [6c_3 + (n-1)(n+2)c_1]x + \sum_{m=2}^{\infty}[(m+2)(m+1)c_{m+2} + (n-m)(n+m+1)c_m]x^m$

$= 0$..............................(5)

Since (5) is an identity , equating the constant term and the coefficient of various powers of x to

zero and obtain $2c_2 + n(n+1)c_0 = 0 \Rightarrow c_2 = -\frac{n(n+1)}{2!}c_0$ ....(6)

$6c_3 + (n-1)(n+2)c_1 = 0 \Rightarrow c_3 = -\frac{(n-1)(n+2)}{3!}c_1$ ....(7)

$[(m+2)(m+1)c_{m+2} + (n-m)(n+m+1)c_m] = 0 \Rightarrow c_{m+2} = -\frac{(n-m)(n+m+1)}{(m+1)(m+2)}c_m;$ for n$\geq$ 2 ....(8)

Putting $m = 2, 3, \ldots$ in (8) and using (6) and (7) , we have

$c_4 = -\frac{(n-2)(n+3)}{4*3}c_2 = \frac{n(n-2)(n+1)(n+3)}{4!}c_0$, $c_5 = -\frac{(n-3)(n+4)}{5*4}c_3 = \frac{(n-3)(n-1)(n+2)(n+4)}{5!}c_1$ ....

Putting these values in (3) , we get

$y = c_0 + c_1 x - \frac{n(n+1)}{2!}c_0 x^2 - \frac{(n-1)(n+2)}{3!}c_1 x^3 + \frac{n(n-2)(n+1)(n+3)}{4!}c_0 x^4 + \frac{(n-3)(n-1)(n+2)(n+4)}{5!}c_1 x^5 + \ldots$

$y = c_0(1 - \frac{n(n+1)}{2!}x^2 + \frac{n(n-2)(n+1)(n+3)}{4!}x^4 - \ldots) + c_1(x - \frac{(n-1)(n+2)}{3!}x^3 + \frac{(n-3)(n-1)(n+2)(n+4)}{5!}x^5 +$

............) ; c_0 and c_1 are arbitrary constants. Which is required solution.

Q-3 State and solve Bessel's differential equation of order P near 0.

The Differential equation of the form $x^2y'' + xy' + (x^2 - P^2)y = 0 \Rightarrow y'' + \frac{1}{x}y' + (1 - \frac{P^2}{x^2})y = 0$

.................(1) is called Bessel's differential equation of order P. Now, solve (1) by using Frobenius method.

Let the solution of (1) be $y = c_0 + c_1x + c_2x^2 + c_3x^3 + \ldots\ldots\ldots$

$y = \sum_{m=0}^{\infty}c_mx^{k+m}$; $c_0 \neq 0$.........................(2) Differentiating (3) twice in succession with respect to 'x' , we get

$y' = \sum_{m=0}^{\infty}c_m(k+m)x^{k+m-1}$ and $y'' = \sum_{m=0}^{\infty}c_m(k+m)(k+m-1)x^{k+m-2}$ ....(4)

Putting the above values of $y, y' and y''$ in (1)

$x^2 \sum_{m=0}^{\infty}(k+m)(k+m-1)c_mx^{k+m-2} + x \sum_{m=0}^{\infty}c_m(k+m)x^{k+m-1} + (x^2 - P^2)\sum_{m=0}^{\infty}c_mx^m = 0$

$\Rightarrow \sum_{m=2}^{\infty}(k+m)(k+m-1)c_mx^{k+m} + \sum_{m=2}^{\infty}(k+m)c_mx^{k+m} - \sum_{m=1}^{\infty}c_mx^{k+m+2} - P^2 \sum_{m=0}^{\infty}c_mx^{k+m} = 0$

$\Rightarrow \sum_{m=2}^{\infty}c_m((k+m)(k+m-1) + (k+m) - P^2) x^{k+m} + \sum_{m=0}^{\infty}c_mx^{k+m+2} = 0$

Equating to zero the smallest power of x, namely x^k , (3) gives the indicial equation

$c_0(k(k-1) + k - p^2) = 0 \Rightarrow c_0(k^2 - p^2) = 0 \Rightarrow k^2 - p^2 = 0; as c_0 \neq 0 \Rightarrow k = p, -p$

Next equating to zero the coefficient of x^{k+1} in (3) gives

$c_1(k(k+1) + (k+1) - p^2) = 0 \Rightarrow c_1((k+1)^2 - p^2) = 0 \Rightarrow c_1((k+1-p)(k+1+p) = 0$

For $k = p, -p$ and $c_1 = 0$, Finally equating to zero the coefficient of x^{k+m} in (3) gives

$c_m((k+m)[(k+m-1) + 1] - p^2) + c_{m-2} = 0 \Rightarrow c_m((k+m+p)((k+m-p))) + c_{m-2} = 0$

$c_m = -\frac{1}{(k+m+p)((k+m-p)}c_{m-2}$ ....(4)

Putting $m = 3,5,7,9,\ldots\ldots\ldots$ in (4) and using $c_1 = 0$ We find $c_1 = c_3 = c_5 = c_7 = 0$

Putting $m = 2,4,6,8,\ldots\ldots\ldots$ in (4) gives $c_2 = -\frac{1}{(k+2+p)((k+2-p)}c_0$

$c_4 = -\frac{1}{(k+4+p)((k+4-p)}c_2 = \frac{1}{(k+4+p)((k+4-p)(k+2+p)((k+2-p)}c_0$ ....

Putting these values in (2) , we get

$y = c_0x^k(1 - \frac{x^2}{(k+2+p)((k+2-p)} - \frac{x^4}{(k+4+p)((k+4-p)(k+2+p)((k+2-p)} - \ldots\ldots\ldots)$

Replacing k by p and c_0 by a the equation gives

$y = ax^p(1 - \frac{x^2}{4((1+p)} + \frac{x^4}{4*8(1+p)((2+p)} - \ldots\ldots\ldots)$ ....(6)

Replacing k by $-p$ and c_0 by b the equation gives

$y = bx^{-p}(1 - \frac{x^2}{4((1+p)} + \frac{x^4}{4*8(1-p)((2-p)} - \ldots\ldots\ldots)$ ....(7)

The particular solution of (1) obtained from (6) by taking arbitrary constant $a = \frac{1}{2^p \Gamma(p+1)}$ is called the Bessel's function of the first kind of order n.

It will be denoted by $J_p(x)$. Thus, we have

$$J_p(x) = \frac{x^p}{2^p \Gamma(p+1)}\left(1 - \frac{x^2}{4((1+p)} + \frac{x^4}{4*8(1+p)((2+p)} - \ldots\ldots)\ldots\ldots\ldots\ldots\ldots\ldots\ldots\ldots(8)$$

$$\Rightarrow J_p(x) = \sum_{r=0}^{\infty}(-1)^r \frac{1}{r!\Gamma(p+r+1)}\left(\frac{x}{2}\right)^{2r+p} \ldots\ldots\ldots\ldots\ldots\ldots\ldots\ldots(9)$$

Replacing b by $\frac{1}{2^p \Gamma(p+1)}$ in (7) , we get

$$J_{-p}(x) = \sum_{r=0}^{\infty}(-1)^r \frac{1}{r!\Gamma(-p+r+1)}\left(\frac{x}{2}\right)^{2r-p} \ldots\ldots\ldots\ldots\ldots\ldots\ldots(10)$$

The general solution of Bessel's equation (1),when p is not an integer is

$$y = AJ_p(x) + BJ_{-p}(x) \; ; \text{ Where } A \text{ and } B \text{ are arbitrary constants.}$$

Q-4 Check the nature of ∞ for $x^4 y'' + x^3(x+2)y' + y = 0$

Given that $x^4 y'' + x^3(x+2)y' + y = 0$.......................(1)

we want to find the solution of (1) for large values of the independent variable.

i.e. about $x = \infty$ for this purpose

we change the independent variable from x to t with the help of the following transformation.

Let $x = \frac{1}{t} \Rightarrow t = \frac{1}{x}$ ....(2)

$$\Rightarrow \frac{dt}{dx} = -\frac{1}{x^2} \ldots\ldots\ldots\ldots\ldots\ldots(3)$$

Now, $y' = \frac{dy}{dx} = \frac{dy}{dt} * \frac{dt}{dx} = \frac{dy}{dt}\left(-\frac{1}{x^2}\right) = -t^2 \frac{dy}{dt}$ ....(4)

$y'' = \frac{d^2y}{dx^2} = \frac{d}{dx}\left(\frac{dy}{dx}\right) = \frac{d}{dt}\left(\frac{dy}{dx}\right)\frac{dt}{dx} = \frac{d}{dt}\left(-\frac{1}{t^2}\frac{dy}{dt}\right)\left(-\frac{1}{x^2}\right) = \left(-t^2\frac{d^2y}{dt^2} - 2t\frac{dy}{dt}\right)(-t^2)$

$$\Rightarrow y'' = t^4 \frac{d^2y}{dt^2} + 2t^3 \frac{dy}{dt} \ldots\ldots\ldots\ldots\ldots\ldots\ldots\ldots(5)$$

Using (3) , (4) and (5) equation (1) transform to $\frac{1}{t^4}\left(t^4\frac{d^2y}{dt^2} + 2t^3\frac{dy}{dt}\right) + \frac{1}{t^3}\left(\frac{1}{t} + 2\right) - t^2\frac{dy}{dt} + y = 0$

$$\Rightarrow \frac{d^2y}{dt^2} + \frac{2}{t}\frac{dy}{dt} - \frac{1}{t^2}\frac{dy}{dt} - \frac{2}{t}\frac{dy}{dt} + y = 0 \Rightarrow \frac{d^2y}{dt^2} - \frac{1}{t^2}\frac{dy}{dt} + y = 0 \ldots\ldots\ldots\ldots\ldots(6)$$

Comparing (6) with $\frac{d^2y}{dt^2} + P(t)\frac{dy}{dt} + Q(t)y = 0$ $P(t) = -\frac{1}{t^2}, Q(t) = 1$

Both are analytic at $t = 0$, So $t = 0$ is a singulur point.

$tP(t) = -\frac{1}{t}, t^2 Q(t) = t^2$ Both are not analytic at $t = 0$,

So, $t = 0$ is an irreguar singulur point of (6)

So, $x = \infty$ is an irreguar singulur point of the given equation.

Q-5 Find an ordinary and singular points of the following differential equation

$$y'' + \frac{x}{(x-1)(x+2)}y' + \frac{1}{(x-1)^2}y = 0$$

Given that $y'' + \frac{x}{(x-1)(x+2)}y' + \frac{1}{(x-1)^2}y = 0$ ....(1)

Comparing (1) with $y'' + P(x)y' + Q(x)y = 0$, Here $P(x) = \frac{x}{(x-1)(x+2)}$, $Q(x) = \frac{1}{(x-1)^2}$

Since $P(x) = \frac{x}{(x-1)(x+2)}$, $Q(x) = \frac{1}{(x-1)^2}$ both are not analytic at $x = 1$ is not an ordinary point of (1) and so $x = 1$ is a singular point.

Now , $P(x), Q(x)$ both are analytic at $x = 2$.

Thus, $x = 2$ is an ordinary point of (1)

So, $x = 1$ is a singular point and is an ordinary point of (1).

Q-6 Discuss the behavior of singular soution of Bessel's equation and find its indicial equation.

The Differential equation of the form $x^2 y'' + xy' + (x^2 - P^2)y = 0$

$\Rightarrow y'' + \frac{1}{x}y' + (1 - \frac{P^2}{x^2})y = 0$ ....(1) is called Bessel's differential equation of order P.

Comparing (1) with $y'' + P(x)y' + Q(x)y = 0$, Here $P(x) = \frac{1}{(x)}$, $Q(x) = \frac{x^2 - p^2}{x^2}$. Since$xP(x) = 1$, $x^2 Q(x) = x^2 - p^2$, both are analytic at $x = 0$ and so $x = 1$ is a regular singular point of (1)

Now, solve (1) by using Frobenius method.

Let the series solution of (1) be of the form

$$y = c_0 + c_1 x + c_2 x^2 + c_3 x^3 + \dots \dots \dots = \sum_{m=0}^{\infty} c_m x^{k+m} \; ; \; c_0 \neq 0 \dots\dots\dots\dots\dots\dots\dots(2)$$

Differentiating (2) twice in succession with respect to 'x' , we get

$y' = \sum_{m=0}^{\infty} c_m(k+m)x^{k+m-1}$ and $y'' = \sum_{m=0}^{\infty} c_m(k+m)(k+m-1)x^{k+m-2}$ ....(4)

Putting the above values of y, y' and y'' in (1)

$x^2 \sum_{m=0}^{\infty}(k+m)(k+m-1)c_m x^{k+m-2} + x \sum_{m=0}^{\infty} c_m(k+m)x^{k+m-1} + (x^2 - P^2) \sum_{m=0}^{\infty} c_m x^m = 0$

$\Rightarrow \sum_{m=2}^{\infty}(k+m)(k+m-1)c_m x^{k+m} + \sum_{m=2}^{\infty}(k+m)c_m x^{k+m} - \sum_{m=1}^{\infty} c_m x^{k+m+2} - P^2 \sum_{m=0}^{\infty} c_m x^{k+m} = 0$

$\Rightarrow \sum_{m=2}^{\infty} c_m((k+m)(k+m-1) + (k+m) - P^2) x^{k+m} + \sum_{m=0}^{\infty} c_m x^{k+m+2} = 0$ or

Equating to zero the smallest power of x, namely x^k , (3) gives the indicial equation

$c_0(k(k-1) + k - p^2) = 0 \Rightarrow c_0(k^2 - p^2) = 0 \Rightarrow k^2 - p^2 = 0; as c_0 \neq 0 \Rightarrow k = p, -p$

Next equating to zero the coefficient of x^{k+1} in (3) gives

$c_1(k(k+1) + (k+1) - p^2) = 0 \Rightarrow c_1((k+1)^2 - p^2) = 0 \Rightarrow c_1((k+1-p)(k+1+p) = 0$

For $k = p, -p$ and $c_1 = 0$, Finally equating to zero the coefficient of x^{k+m} in (3) gives

$c_m((k+m)[(k+m-1)+1] - p^2) + c_{m-2} = 0 \Rightarrow c_m((k+m+p)((k+m-p))) + c_{m-2} = 0$

Q-7 Derive an integral representation for the Gauss Hyper geometric function.

$F(\alpha, \beta; \gamma; x) = \frac{\Gamma(\gamma)}{\Gamma\beta\Gamma(\gamma-\beta)} \int_0^t t^{\beta-1}(1-t)^{\gamma-\beta-1}(1-xt)^{-\alpha}dt$ or

$F(\alpha, \beta; \gamma; x) = \frac{1}{\beta(\beta,\gamma-\beta)} \int_0^t t^{\beta-1}(1-t)^{\gamma-\beta-1}(1-xt)^{-\alpha}dt$; if $\gamma > \beta > 0$

By definition we have $F(\alpha, \beta; \gamma; x) = \sum_{n=1}^{\infty} \frac{(\alpha)_n(\beta)_n}{(\gamma)_n}\frac{x^n}{n!} = \sum_{n=1}^{\infty}(\alpha)_n \frac{\Gamma(\beta+n)}{\beta}\frac{\Gamma(\gamma)}{\Gamma(\gamma+n)}\frac{x^n}{n!}$

Multiplying and dividing by $\Gamma(\gamma - \beta)$

$F(\alpha, \beta; \gamma; x) = \frac{\Gamma(\gamma-\beta)}{(\Gamma(\gamma-\beta))} \sum_{n=1}^{\infty}(\alpha)_n \frac{\Gamma(\beta+n)}{\beta}\frac{\Gamma(\gamma)}{\Gamma(\gamma+n)}\frac{x^n}{n!}$

$F(\alpha, \beta; \gamma; x) = \frac{\Gamma\gamma}{(\Gamma(\gamma-\beta))\Gamma\beta} \sum_{n=1}^{\infty}(\alpha)_n \frac{\Gamma(\beta+n)\Gamma(\gamma-\beta)}{\Gamma(\beta+n+\gamma-\beta)}\frac{x^n}{n!} = \frac{\Gamma\gamma}{(\Gamma(\gamma-\beta))\Gamma\beta} \sum_{n=1}^{\infty}(\alpha)_n\beta(\beta+n, \gamma-\beta)\frac{x^n}{n!}$

Now, $\beta(m, n) = \int_0^t t^{m-1}(1-t)^{n-1}dt$

$F(\alpha, \beta; \gamma; x) = \frac{\Gamma(\gamma)}{\Gamma\beta\Gamma(\gamma-\beta)} \sum_{n=1}^{\infty}(\alpha)_n\int_0^t t^{\beta+n-1}(1-t)^{\gamma-\beta-1}dt\frac{x^n}{n!}$

Now, $(1-xt)^{-\alpha} = \frac{(-\alpha)(-\alpha-1)\ldots\ldots\ldots(-\alpha-n+1)}{n!}(-xt)^n = (-1)^n\frac{\alpha(\alpha+n)(\alpha+n-1)}{n!}(-1)^nx^nt^n = (\alpha)_n\frac{x^nt^n}{n!}$

$F(\alpha, \beta; \gamma; x) = \frac{\Gamma(\gamma)}{\Gamma\beta\Gamma(\gamma-\beta)} \int_0^t t^{\beta-1}(1-t)^{\gamma-\beta-1}(1-xt)^{-\alpha}dt$

Also, $\beta(\beta, \gamma-\beta) = \frac{\Gamma\beta\Gamma(\gamma-\beta)}{\Gamma(\beta+\gamma-\beta)} = \frac{\Gamma\beta\Gamma(\gamma-\beta)}{\Gamma\gamma} \Rightarrow \frac{1}{\beta(\beta,\gamma-\beta)} = \frac{\Gamma\gamma}{\Gamma\beta\Gamma(\gamma-\beta)}$

Hence, $F(\alpha, \beta; \gamma; x) = \frac{1}{\beta(\beta,\gamma-\beta)} \int_0^t t^{\beta-1}(1-t)^{\gamma-\beta-1}(1-xt)^{-\alpha}dt$; if $\gamma > \beta > 0$

Q-8 Define regular singular point. Find Frobenius series expansion $2x^2y'' + x(2x+1)y' - y = 0$ near regular singular point.

A singular point $x = x_0$ of the differential equation $y'' + P(x)y' + Q(x)y = 0$ is called a regular singular point of the differential equation $y'' + P(x)y' + Q(x)y = 0$ if both $(x - x_0)P(x)$ and $(x - x_0)^2Q(x)$ are analytic at $x = x_0$. For Frobenius series expansion

Given equation $2x^2y'' + x(2x+1)y' - y = 0$ ....(1)

Dividing by $2x^2$, the given equation becomes $y'' + \frac{(2x+1)}{2x}y' - \frac{1}{2x^2}y = 0$ ....(1)

Comparing (1) with standard equation $y'' + P(x)y' + Q(x)y = 0$ if both $(x - x_0)P(x)$,

we have $P(x) = \frac{(2x+1)}{2x}$ and $Q(x) = -\frac{1}{2x^2}$

Since P(x) and Q(x) are not both analytic at $x = 0$. So,$x = 0$ is not ordinary point

Now, $xP(x) = \frac{2x+1}{2}$, $x^2Q(x) = -\frac{1}{2}$

Showing that both $P(x)$ and $Q(x)$ are analytic at $x = 0$ is a regular singular point.

To find series expansion , we take Let the solution of (1) be

$$y = c_0 + c_1 x + c_2 x^2 + c_3 x^3 + \ldots\ldots\ldots = \sum_{m=0}^{\infty} c_m x^{k+m} \; ; \; c_0 \neq 0 \ldots\ldots\ldots\ldots\ldots\ldots(2)$$

Differentiating (3) twice in succession with respect to 'x' , we get

$$y' = \sum_{m=0}^{\infty} c_m (k+m) x^{k+m-1} \text{ and } y'' = \sum_{m=0}^{\infty} c_m (k+m)(k+m-1) x^{k+m-2} \ldots\ldots\ldots\ldots\ldots\ldots(4)$$

Putting the above values of y, y' and y'' in (1)

$$2x^2 \sum_{m=0}^{\infty} (k+m)(k+m-1) c_m x^{k+m-2} + x(2x+1) \sum_{m=0}^{\infty} c_m (k+m) x^{k+m-1} + \sum_{m=0}^{\infty} c_m x^m = 0$$

$$\Rightarrow 2 \sum_{m=0}^{\infty} (k+m)(k+m-1) c_m x^{k+m} + 2 \sum_{m=0}^{\infty} c_m (k+m) x^{k+m+1} + \sum_{m=0}^{\infty} c_m (k+m) x^{k+m} \sum_{m=0}^{\infty} c_m x^{k+m} = 0$$

$$\sum_{m=0}^{\infty} c_m [(2(k+m)(k+m-1)) + (k+m) - 1] x^{k+m} + \sum_{m=0}^{\infty} c_m [k+m] x^{k+m+1} = 0 \ldots\ldots\ldots\ldots(4)$$

Which is an identity equating to zero the coefficient of the smallest power in x, namely x^k gives

the indicial equation.

$$c_0(2k(k-1) + k - 1) = 0 \Rightarrow c_0(2k^2 - k - 1) = 0 \Rightarrow 2k^2 - k - 1 = 0 \text{ as } c_0 \neq 0$$

$$\Rightarrow (k-1)(2k+1) = 0 \Rightarrow k = 1, -\tfrac{1}{2} \ldots\ldots\ldots\ldots(5) \text{ Now we equate to zero coefficient of } x^{k+m}$$

$$c_m[2(k+m)(k+m-1) + (k+m) - 1] + c_{m-1}(k+m-1) = 0$$

$$\Rightarrow c_m = -\frac{(k+m-1)}{[2(k+m)(k+m-1)+(k+m)-1]} c_{m-1} \ldots\ldots\ldots\ldots\ldots\ldots(6)$$

Taking $m = 1, 2, 3, \ldots\ldots\ldots$ in (6), gives $c_1 = -\frac{k}{(k+1)(2k+1)-1} c_0$

$$c_2 = -\frac{(k+1)}{(k+2)(2k+3)-1} c_1 \Rightarrow c_2 = -\frac{k(k+1)}{[(k+2)(2k+3)-1][(k+1)(2k+1)-1]} c_0 \ldots\ldots\ldots\ldots\ldots\ldots$$

Put these values in (2) $y = x^k [c_0 + c_1 x + c_2 x^2 + c_3 x^3 + \ldots\ldots\ldots]$

$$\Rightarrow y = x^k [c_0 + [-\frac{k}{(k+1)(2k+1)-1} c_0] x + [-\frac{k(k+1)}{[(k+2)(2k+3)-1][(k+1)(2k+1)-1]}] c_0 x^2 + \ldots\ldots\ldots]$$

$$\Rightarrow y = c_0 x^k [1 + [-\frac{k}{(k+1)(2k+1)-1}] x + [-\frac{k(k+1)}{[(k+2)(2k+3)-1][(k+1)(2k+1)-1]}] x^2 + \ldots\ldots\ldots] \ldots\ldots\ldots\ldots\ldots\ldots(7)$$

Putting $k = 1$ and replacing c_0 by a in (7),

$$y = ax [1 + [-\frac{1}{(2)(3)-1}] x + -\frac{k(k+1)}{[(k+2)(2k+3)-1][(k+1)(2k+1)-1]} c_0 x^2 + \ldots\ldots\ldots]$$

$$\Rightarrow y = ax [1 + [-\frac{1}{(2)(3)-1}] x + [-\frac{(1)(2)}{[(3)(5)-1][(2)(3)-1]}] x^2 + \ldots\ldots\ldots]$$

$$\Rightarrow y = ax [1 - \tfrac{1}{5} x + \tfrac{2}{70} x^2 + \ldots\ldots\ldots\ldots] = au$$

Putting $k = -\tfrac{1}{2}$ and replacing c_0 by b in (7)

$$y = bx^{-\frac{1}{2}} [1 - [-\frac{-\frac{1}{2}}{(-\frac{1}{2}+1)(-1+1)-1}] x + [-\frac{-\frac{1}{2}(\frac{1}{2})}{[(-\frac{1}{2}+2)(-1+3)-1][(\frac{1}{2}+1)(0)-1]}] x^2 + \ldots\ldots\ldots]$$

$$\Rightarrow y = bx^{-\frac{1}{2}} [1 + \tfrac{1}{2} x + \tfrac{1}{4} x^2 + \ldots\ldots\ldots] = bv$$

So, the required general solution is $y = au + bv$

$$y = ax [1 - \tfrac{1}{5} x + \tfrac{2}{70} x^2 + \ldots\ldots\ldots\ldots] + bx^{-\frac{1}{2}} [1 + \tfrac{1}{2} x + \tfrac{1}{4} x^2 + \ldots\ldots\ldots]$$

Q-9 State only Rodrigue's formula and find Legendre polynomials $P_0(x)$,$P_1(x)$,$P_2(x)$ and $P_3(x)$. Also express $x^3 + 5x + 2$ in terms of Legendre polynomials.

Rodrigue's formula $P_n(x) = \frac{1}{2^n n!} \frac{d^n}{dx^n}(x^2-1)^n$ ....(1)

Let $n = 0$ in (1) $P_0(x) = \frac{1}{2^0 0!} \frac{d^0}{dx^0}(x^2-1)^0 = 1$ ....(2)

Let $n = 1$ in (1) $P_1(x) = \frac{1}{2^1 1!} \frac{d^1}{dx^1}(x^2-1)^1 = x$ ....(3)

Let $n = 2$ in (1) $P_2(x) = \frac{1}{2^2 2!} \frac{d^2}{dx^2}(x^2-1)^2 = \frac{1}{2}(3x^2-1)$ ....(4)

Let $n = 3$ in (1) $P_3(x) = \frac{1}{2^3 3!} \frac{d^3}{dx^3}(x^2-1)^3 = \frac{1}{2}(5x^3-3x)$ ....(5)

From (2) , (3) , (4) and (5) gives $x^3 = \frac{2P_3(x)+3x}{5}$ and $x^2 = \frac{2P_2(x)+1}{3}$

Hence , $x^3 + 5x + 2 = \frac{1}{5}(2P_3(x) + 28P_1(x) + 10P_0(x))$

Q-10 State and prove orthogonality of Legendre polynomials.

or

Prove that $\int_{-1}^{1} P_m(x)P_n(x)dx = \begin{cases} 0 & \text{if } m \neq n \\ \\ \frac{2}{2n+1} & \text{if } m = n \end{cases}$

When $m \neq n$ Since $P_m(x)$ and $P_n(x)$ satisfied Legendre's equation, we have

$(1-x^2)P_m''(x) - 2xP_m'(x) + m(m+1)P_m(x) = 0$ ....(1)

$(1-x^2)P_n''(x) - 2xP_n'(x) + n(n+1)P_n(x) = 0$ ....(2)

Multiplying (1) by $P_n(x)$ and (2) by $P_m(x)$ and then have subtracting the resulting equation , we have

$(1-x^2)(P_n(x)P_m''(x) - P_m(x)P_n''(x)) - 2x(P_n(x)P_m'(x) - P_m(x)P_n'(x)) + (m(m+1) - n(n+1))P_m(x)P_n(x) = 0$

$\Rightarrow (1-x^2)(P_nP_m'' - P_mP_n'') - 2x(P_nP_m' - P_mP_n') + (m(m+1) - n(n+1))P_mP_n = 0$

$\Rightarrow \frac{d}{dx}[(1-x^2)(P_nP_m'' - P_mP_n'')] + [m(m+1) - n(n+1)]P_mP_n = 0$

$\Rightarrow \frac{d}{dx}[(1-x^2)(P_nP_m'' - P_mP_n'')] = [(n-m)(n+m+1)]P_mP_n$

Integrating both sides with respect to x from -1 to 1 , we get

$(n-m)(n+m+1) \int_{-1}^{1} P_m(x)P_n(x)dx = [(1-x^2)(P_nP_m' - P_mP_n')]_{x=-1}^{x=1}$

So, $\int_{-1}^{1} P_m(x)P_n(x)dx = 0$ as $m \neq n$ ....(4)

Page 8

When $m = n$

We know that Generating Function

$(1-2xz+z^2)^{-\frac{1}{2}} = \sum_{n=0}^{\infty} Z^n P_n(x)$ ....(5) and $(1-2xz+z^2)^{-\frac{1}{2}} = \sum_{m=0}^{\infty} Z^m P_m(x)$ ....(6)

Multiplying (5) and (6) , we get $\sum_{n=0}^{\infty} \sum_{m=0}^{\infty} P_n(x)P_m(x)Z^{n+m} = (1 - 2xz + z^2)^{-1}$

Integrating both sides with respect to x from -1 to 1 , we get

$\sum_{n=0}^{\infty} \sum_{m=0}^{\infty} (\int_{-1}^{1} P_n(x)P_m(x)dx)Z^{n+m} = \int_{-1}^{1}(1 - 2xz + z^2)^{-1}dx$ ....(7)

Now from (4) and (7)

$\sum_{n=0}^{\infty} \int_{-1}^{1}((P_n(x))^2dx)Z^{2n} = \int_{-1}^{1} \frac{dx}{1-2xz+z^2} = [\frac{log(1-2xz+z^2)}{-2z}]_{-1}^{1} = -\frac{1}{2z}[log(1 - z)^2 - log(1 + z)^2]$

$= \frac{1}{z}[log(1 + z) - log(1 - z)] = \frac{2}{z}(z + \frac{z^3}{3} + \frac{z^5}{5}) = \frac{2}{z}\sum_{n=0}^{\infty} \frac{z^{2n+1}}{2n+1}$

Equating coefficient z^{2n} ,we get $\sum_{n=0}^{\infty} \int_{-1}^{1}((P_n(x))^2dx) = \frac{2}{2n+1}$ ....(8)

Combining (4) and (8) , we get $\int_{-1}^{1} P_m(x)P_n(x)dx = \begin{cases} 0 & \text{if } m \neq n \\ \\ \frac{2}{2n+1} & \text{if } m = n \end{cases}$

Q-11 State and prove orthogonality of Bessel's function.

If λ_i and λ_j are roots of the equation $J_n(\lambda a) = 0$ then $\int_0^a xJ_n(\lambda_i x)J_n(\lambda_j x)dx = \frac{a^2}{2} J_{n+1}^2(\lambda_i a)\delta_{ij}$

Case : I

Let $i \neq j$ i.e. let λ_i and λ_j be unequal roots of $J_n(\lambda a) = 0$

$J_n(\lambda_i a) = 0$ and $J_n(\lambda_j a) = 0$ ....(1)

Let $u(x) = J_n(\lambda_i x)$ and $v(x) = J_n(\lambda_j x)$ ....(2)

then u and v are Bessel functions satisfied modified Bessel's equation

$x^2y'' + xy' + (\lambda^2x^2 - n^2)y = 0$ ....(3) then we have

$x^2u'' + xu' + (\lambda_i^2x^2 - n^2)u = 0$ ....(4)

$x^2v'' + xv' + (\lambda_j^2x^2 - n^2)v = 0$ ....(5)

Multiplying (4) by v and (5) by u and then subtracting , we get

$x^2(vu'' - u v'') + x(vu' - u v') + x^2(\lambda_i^2 - \lambda_j^2)uv = 0$

$x^2(vu'' - u\,v'') + x(vu' - u\,v') = x^2(\lambda_j^2 - \lambda_i^2)uv$

$x(vu'' - u\,v'') + (vu' - u\,v') = x(\lambda_j^2 - \lambda_i^2)uv \quad \frac{d}{dx}(x(vu' - u\,v')) = x(\lambda_j^2 - \lambda_i^2)uv$ ....(6)

Integrating (6) with respect to x from 0 to a ,

$(\lambda_j^2 - \lambda_i^2)\int_0^a xuv\,dx = [(x(vu' - u\,v'))]_0^a$ ....(7)

Using (2) , (7) reduce to

$(\lambda_j^2 - \lambda_i^2)\int_0^a xJ_n(\lambda_i x)J_n(\lambda_j x)dx = [(x(J_n(\lambda_j x)J'_n(\lambda_i x) - J_n(\lambda_i x)J'_n(\lambda_j x)))]_0^a$ ....(7)

By using (1)

$(\lambda_j^2 - \lambda_i^2)\int_0^a xJ_n(\lambda_i x)J_n(\lambda_j x)dx = 0$ Since $\lambda_i \neq \lambda_j$ the above equation gives

$\int_0^a xJ_n(\lambda_i x)J_n(\lambda_j x)dx = 0$, when $i \neq j$ ....(8)

Case II

Let $i = j$ (equal roots) Multiplying (4) by $2u'$, we have

$2x^2u''\,u' + 2xu' + 2(\lambda_i^2 x^2 - n^2)uu' = 0 \Rightarrow 2x^2u''\,u' + 2xu' - 2n^2uu' + 2\lambda_i^2 x^2uu' = 0$

$\Rightarrow 2x^2u''\,u' + 2xu' - 2n^2uu' + 2\lambda_i^2 x^2uu' + 2\lambda_i^2 xu^2 - 2\lambda_i^2 xu^2 = 0$

$\Rightarrow \frac{d}{dx}(xu'^2 - n^2u^2 + \lambda_i^2 x^2u^2) - 2\lambda_i^2 xu^2 = 0$

$\Rightarrow 2\lambda_i^2 xu^2 = \frac{d}{dx}(xu'^2 - n^2u^2 + \lambda_i^2 x^2u^2)$

Integrating (9) with respect to x from 0 to a,

$2\lambda_i^2 \int_0^a xu^2 dx = [xu'^2 - n^2u^2 + \lambda_i^2 x^2u^2]_0^a$ ....(10)

Using (1) , (2) and $J_n(0) = 0$, we have $2\lambda_i^2 \int_0^a xJ_n^2(\lambda_i(x))dx = a^2[J'_n(\lambda_i(x))^2]_{x=a}$ ....(11)

from reccurrence relation $\frac{d}{dx}J_n(x) = \frac{n}{x}J_n(x) - J_{n+1}(x)$ ....(12)

Replacing x by $\lambda_i(x)$ in (12) , we have $\frac{d}{d(\lambda_i x)}J_n(\lambda x) = \frac{n}{\lambda_i x}J_n(\lambda_i x) - J_{n+1}(\lambda_i x)$

$\Rightarrow \frac{1}{\lambda_i}\frac{d}{dx}J_n(\lambda_i x) = \frac{n}{\lambda_i x}J_n(\lambda_i x) - J_{n+1}(\lambda_i x)$

$\Rightarrow J'_n(\lambda_i x) = \frac{n}{x}J_n(\lambda_i x) - \lambda_i J_{n+1}(\lambda_i x)$

$\Rightarrow [J'_n(\lambda_i x)^2]_{x=a} = [\frac{n}{x}J_n(\lambda_i x) - \lambda_i J_{n+1}(\lambda_i x)^2]_{x=a} = [0 - \lambda_i J_{n+1}(\lambda_i x)^2]$ ....by(1)

$\Rightarrow [J'_n(\lambda_i x)^2]_{x=a} = \lambda_i^2 J_{n+1}^2(\lambda_i a)^2$

Using this value in (11) and dividing both sides of the resulting equation by $2\lambda_i^2$, we get

$\int_0^a xJ_n^2(\lambda_i x)dx = \frac{a^2}{2}J_{n+1}(\lambda_i a)$ ....(10) Combining (8) and (13) , we have

$\int_0^a xJ_n(\lambda_i x)J_n(\lambda_j x)dx = \frac{a^2}{2}J_{n+1}(\lambda_i a)\delta_{ij}$

Q-12   (I) Prove that $(n+1)P_{n+1} = (2n+1)xP_n - nP_{n-1}$

We know that generating function $(1-2xz+z^2)^{-\frac{1}{2}} = \sum_{n=0}^{\infty} Z^n P_n(x)$ ....(1)

Differentiating both sides with respect to z of (1)

$-\frac{1}{2}(1-2xz+z^2)^{-\frac{3}{2}}(-2x+2z) = \sum_{n=0}^{\infty} nZ^{n-1}P_n(x)$ ....(2)

$\Rightarrow (1-2xz+z^2)^{-\frac{3}{2}}(x-z) = \sum_{n=0}^{\infty} nZ^{n-1}P_n(x)$

Multiplying both sides by $(1-2xz+z^2)$

$(x-z)(1-2xz+z^2)^{-\frac{1}{2}} = (1-2xz+z^2)\sum_{n=0}^{\infty} nZ^{n-1}P_n(x)$, Now from (1)

$(x-z)\sum_{n=0}^{\infty} Z^n P_n(x) = (1-2xz+z^2)\sum_{n=0}^{\infty} nZ^{n-1}P_n(x)$

$\Rightarrow x\sum_{n=0}^{\infty} Z^n P_n(x) - z\sum_{n=0}^{\infty} Z^n P_n(x)$

$= \sum_{n=0}^{\infty} nZ^{n-1}P_n(x) - 2xz\sum_{n=0}^{\infty} nZ^{n-1}P_n(x) + z^2\sum_{n=0}^{\infty} nZ^{n-1}P_n(x)$

$\Rightarrow x\sum_{n=0}^{\infty} Z^n P_n(x) - \sum_{n=0}^{\infty} Z^{n+1}P_n(x)$

$= \sum_{n=0}^{\infty} nZ^{n-1}P_n(x) - 2x\sum_{n=0}^{\infty} nZ^n P_n(x) + \sum_{n=0}^{\infty} nZ^{n+1}P_n(x)$

$\Rightarrow x\sum_{n=0}^{\infty} Z^n P_n(x) - \sum_{n=0}^{\infty} Z^n P_{n-1}(x)$

$= \sum_{n=0}^{\infty}(n+1)Z^n P_{n+1}(x) - 2x\sum_{n=0}^{\infty} nZ^n P_n(x) + \sum_{n=0}^{\infty}(n-1)Z^{n+1}P_{n-1}(x)$

Equating coefficient of z^n on both sides , we get

$xP_n(x) - P_{n-1}(x) = (n+1)P_{n+1}(x) - 2xnP_n(x) + (n-1)P_{n-1}(x)$

$(n+1)P_{n+1}(x) = xP_n(x) - P_{n-1}(x) + 2xnP_n(x) - nP_{n-1}(x) + P_{n-1}(x)$

$(n+1)P_{n+1}(x) = (2n+1)xP_n(x) - nP_{n-1}(x)$

(II) Prove that $nP_n = (2n+1)xP_n' - P_{n-1}'$

We have generating function $(1-2xz+z^2)^{-\frac{1}{2}} = \sum_{n=0}^{\infty} Z^n P_n(x)$ ....(1)

Differentiating both sides with respect to z of (1)

$-\frac{1}{2}(1-2xz+z^2)^{-\frac{3}{2}}(-2x+2z) = \sum_{n=0}^{\infty} nZ^{n-1}P_n(x)$ ....(2)

$\Rightarrow (1-2xz+z^2)^{-\frac{3}{2}}(x-z) = \sum_{n=0}^{\infty} nZ^{n-1}P_n(x)$ ....(3)

Again differentiating with respect to x of (1), we get

$(1-2xz+z^2)^{-\frac{3}{2}}(-2z) = \sum_{n=0}^{\infty} nZ^n P_n'(x)$

$\Rightarrow z(1-2xz+z^2)^{-\frac{3}{2}} = \sum_{n=0}^{\infty} nZ^n P_n'(x)$

Multiplying both sides by $(x-z)$, we get

$z(x-z)(1-2xz+z^2)^{-\frac{3}{2}} = (x-z)\sum_{n=0}^{\infty} nZ^n P_n'(x)$

Now from (3)

$z\sum_{n=0}^{\infty} nZ^{n-1}P_n(x) = (x-z)\sum_{n=0}^{\infty} nZ^n P_n'(x)$

$$\sum_{n=0}^{\infty} n Z^n P_n(x) = x \sum_{n=0}^{\infty} Z^n P_n'(x) - z \sum_{n=0}^{\infty} Z^n P_n'(x)$$

$$\sum_{n=0}^{\infty} n Z^n P_n(x) = x \sum_{n=0}^{\infty} Z^n P_n'(x) - \sum_{n=0}^{\infty} Z^n P_{n-1}'(x)$$ Now equating coefficient of z^n on both sides , we get

$$n P_n = (2n+1) x P_n' - P_{n-1}'$$

Q-13 Prove that $\frac{d}{dx}[x^n J_n(x)] = x^n J_{n-1}(x)$

We know that $J_n(x) = \sum_{r=0}^{\infty} (-1)^r \frac{1}{r! \Gamma(n+r+1)} \left(\frac{x}{2}\right)^{2r+n}$

Multiplying both sides by x^n and then taking derivative with respect to x

$$\frac{d}{dx}[x^n J_n(x)] = \frac{d}{dx}\left[\sum_{r=0}^{\infty} (-1)^r \frac{1}{r! \Gamma(n+r+1)} \left(\frac{x}{2}\right)^{2r+n}\right]$$

$$\Rightarrow \frac{d}{dx}[x^n J_n(x)] = \sum_{r=0}^{\infty} (-1)^r \frac{1}{r! \Gamma(n+r+1)} \left(\frac{1}{2}\right)^{2r+n} \frac{d}{dx}[x^{2r+n}]$$

$$\Rightarrow \frac{d}{dx}[x^n J_n(x)] = \sum_{r=0}^{\infty} \frac{(-1)^r}{r!} \frac{2r+2n}{\Gamma(n+r+1)} \frac{x^{2r+2n-1}}{2^{2r+n}}$$

$$\Rightarrow \frac{d}{dx}[x^n J_n(x)] = \sum_{r=0}^{\infty} \frac{(-1)^r}{r!} \frac{2(r+n)x^n}{\Gamma(n+r+1)} \frac{x^{2r+n-1}}{2^{2r+n}}$$

$$\Rightarrow \frac{d}{dx}[x^n J_n(x)] = \sum_{r=0}^{\infty} \frac{(-1)^r 2(r+n)x^n}{r!(n+r)\Gamma(n+r)} \frac{x^{2r+n-1}}{2^{2r+n}}$$

$$\Rightarrow \frac{d}{dx}[x^n J_n(x)] = x^n \sum_{r=0}^{\infty} \frac{(-1)^r}{r! \Gamma(n-1+r+1)} \left(\frac{x}{2}\right)^{2r+n-1} = x^n J_{n-1}(x)$$

So, $\frac{d}{dx}[x^n J_n(x)] = x^n J_{n-1}(x)$

Q-14 Prove that $\frac{d}{dx}[x^{-n} J_n(x)] = -x^{-n} J_{n+1}(x)$

We know that $J_n(x) = \sum_{r=0}^{\infty} (-1)^r \frac{1}{r! \Gamma(n+r+1)} \left(\frac{x}{2}\right)^{2r+n}$

Multiplying both sides by x^{-n} and then taking derivative with respect to x

$$\frac{d}{dx}[x^{-n} J_n(x)] = \frac{d}{dx}\left[x^{-n} \sum_{r=0}^{\infty} (-1)^r \frac{1}{r! \Gamma(n+r+1)} \left(\frac{x}{2}\right)^{2r+n}\right]$$

$$\Rightarrow \frac{d}{dx}[x^{-n} J_n(x)] = \sum_{r=0}^{\infty} (-1)^r \frac{1}{r! \Gamma(n+r+1)} \left(\frac{1}{2}\right)^{2r+n} \frac{d}{dx}[x^{2r}]$$

$$\Rightarrow \frac{d}{dx}[x^{-n} J_n(x)] = \sum_{r=0}^{\infty} \frac{(-1)^r 2r x^{2r-1}}{r(r-1)! \Gamma(n+r+1) 2^{2r+n}}$$

$$\Rightarrow \frac{d}{dx}[x^{-n} J_n(x)] = \sum_{r=1}^{\infty} \frac{(-1)^r 2r x^{2r-1}}{(r-1)! \Gamma(n+r+1) 2^{2r+n-1}}$$

(Because when $r = 0$ corresponding term vanishes) Now take $m = r - 1 \Rightarrow r = m+1$

If $r = 1 \Rightarrow m = 0$ and $r = \infty \Rightarrow m = \infty$

$$\Rightarrow \frac{d}{dx}[x^{-n} J_n(x)] = \sum_{m=0}^{\infty} \frac{(-1)^{m+1} x^{2m+2-1}}{m! \Gamma(n+m+2)} \frac{x^n x^{-n}}{2^{2m+2+n-1}}$$

$$\Rightarrow \frac{d}{dx}[x^{-n} J_n(x)] = -x^{-n} \sum_{m=0}^{\infty} \frac{(-1)^{m+1} x^{2m+2-1}}{m! \Gamma(n+m+2)} \left(\frac{x}{2}\right)^{2m+n+1}$$

On changing the variable of summation from m to r

$$\Rightarrow \frac{d}{dx}[x^{-n} J_n(x)] = -x^{-n} \sum_{r=0}^{\infty} \frac{(-1)^r}{r! \Gamma(n+r+2)} \left(\frac{x}{2}\right)^{2r+n+1}$$

Hence $\frac{d}{dx}[x^{-n} J_n(x)] = -x^{-n} J_{n+1}(x)$

Q-15 State and prove Rodrigue's formula for the Legendre polynomial and hence express $x^2 - 3x + 1$ in terms of Legendre polynomials.

$P_n(x) = \frac{1}{2^n n!}\frac{d^n}{dx^n}(x^2 - 1)^n$ By definition of Legendre polynomial

$P_n(x) = \sum_{r=0}^{[\frac{n}{2}]}(-1)^r \frac{(2n-2r)!x^{n-2r}}{2^n r!(n-r)!(n-2r)!}$; where $[\frac{n}{2}] = \begin{cases} \frac{n}{2} & \text{; if } n \text{ is } even \\ \frac{n-1}{2} & \text{; if } n \text{ is } odd \end{cases}$

By Bionomial theorem $(x^2 - 1)^n = \sum_{r=0}^{n}\binom{n}{r}(x^2)^{n-r}(-1)^r = \sum_{r=0}^{n}\binom{n}{r}(-1)^r(x)^{2n-2r}$

So, $\frac{d^n}{dx^n}(x^2 - 1)^n = \sum_{r=0}^{n}\binom{n}{r}(-1)^r\frac{d^n}{dx^n}(x)^{2n-2r}$

$\Rightarrow \frac{1}{2^n n!}\frac{d^n}{dx^n}(x^2 - 1)^n = \frac{1}{2^n n!}\sum_{r=0}^{n}\binom{n}{r}(-1)^r\frac{d^n}{dx^n}(x)^{2n-2r}$ ....(1)

But $\frac{d^n}{dx^n}(x)^m = 0$; if $m < n$ and $\frac{d^n}{dx^n}(x)^m = \frac{m!}{(m-n)!}x^{m-n}$ if $m \geq n$ ....(2)

$\therefore \frac{d^n}{dx^n}(x^{2n-2r}) = 0$; if $2n - 2r < n$ i.e. $r > \frac{n}{2}$ ....(3)

From (3) and (2) we must replace $\sum_{r=0}^{n}$ by $\sum_{r=0}^{[\frac{n}{2}]}$; if n is even and $\sum_{r=0}^{(\frac{n-1}{2})}$; if n is odd i.e. we must replace to $\sum_{r=0}^{n}$ by $\sum_{r=0}^{[\frac{n}{2}]}$

Hence (1) reduces to $\frac{1}{2^n n!}\frac{d^n}{dx^n}(x^2 - 1)^n = \frac{1}{2^n n!}\sum_{r=0}^{[\frac{n}{2}]}\binom{n}{r}(-1)^r\frac{d^n}{dx^n}(x)^{2n-2r}$ Now from (2)

$\frac{1}{2^n n!}\frac{d^n}{dx^n}(x^2 - 1)^n = \frac{1}{2^n n!}\sum_{r=0}^{[\frac{n}{2}]}\binom{n}{r}(-1)^r\frac{(2n-2r)!}{(2n-2r-n)!}(x)^{2n-2r-n}$

$\therefore \frac{1}{2^n n!}\frac{d^n}{dx^n}(x^2 - 1)^n = \frac{1}{2^n n!}\sum_{r=0}^{[\frac{n}{2}]}\frac{(-1)^r n!}{r!(n-r)!}\binom{n}{r}\frac{(2n-2r)!}{(n-2r)!}(x)^{n-2r}$

$\therefore \frac{1}{2^n n!}\frac{d^n}{dx^n}(x^2 - 1)^n = \sum_{r=0}^{[\frac{n}{2}]}(-1)^r\frac{(2n-2r)!}{(2n-2r)!}\binom{n}{r}\frac{(2n-2r)!}{(n-2r)!}(x)^{n-2r} = P_n(x)$ ....(3)

If $n = 0$ then $P_0(x) = 1$, If $n = 1$ then $P_1(x) = x$ and If $n = 2$ then

$P_2(x) = \frac{3x^2 - 1}{2} \Rightarrow x^2 = \frac{2P_2(x)+1}{3}$

So,$x^2 - 3x + 1 = \frac{1}{3}[3P_0(x) - 9P_1(x) + 2P_2(x) + 1]$

Q-16 In usual notation prove that $\int_{-1}^{1} P_m(x)P_n(x)dx = \begin{cases} 0 & \text{; if } m \neq n \\ \frac{2}{2n+1} & \text{; if } m = n \end{cases}$

When m Since $P_m(x)$ and $P_n(x)$ satisfied Legendre's equation, we have

$(1 - x^2)P_m''(x) - 2xP_m'(x) + m(m + 1)P_m(x) = 0$ ....(1)

$(1 - x^2)P_n''(x) - 2xP_n'(x) + n(n + 1)P_n(x) = 0$ ....(2)

Multiplying (1) by $P_n(x)$ and (2) by $P_m(x)$ and then have subtracting the resulting equation , we have

$(1 - x^2)(P_n(x)P_m''(x) - P_m(x)P_n''(x)) - 2x(P_n(x)P_m'(x) - P_m(x)P_n'(x)) + (m(m + 1) - n(n + 1))P_m(x)P_n(x) = 0 \Rightarrow (1 - x^2)(P_nP_m'' - P_mP_n'') - 2x(P_nP_m' - P_mP_n') + (m(m + 1) - n(n + 1))P_mP_n = 0$

$\Rightarrow \frac{d}{dx}[(1-x^2)(P_n P_m'' - P_m P_n'')] + [m(m+1) - n(n+1)]P_m P_n = 0$

$\Rightarrow \frac{d}{dx}[(1-x^2)(P_n P_m'' - P_m P_n'')] = [(n-m)(n+m+1)]P_m P_n$

Integrating both sides with respect to x from -1 to 1 , we get

$(n-m)(n+m+1) \int_{-1}^{1} P_m(x)P_n(x)dx = [(1-x^2)(P_n P_m' - P_m P_n')]_{x=-1}^{x=1}$

So, $\int_{-1}^{1} P_m(x)P_n(x)dx = 0$ as $m \neq n$ ....(4)

When $m = n$

We know that Generating Function $(1-2xz+z^2)^{-\frac{1}{2}} = \sum_{n=0}^{\infty} Z^n P_n(x)$ ....

and $(1-2xz+z^2)^{-\frac{1}{2}} = \sum_{m=0}^{\infty} Z^m P_m(x)$ ....(6)

Multiplying (5) and (6) , we get $\sum_{n=0}^{\infty} \sum_{m=0}^{\infty} P_n(x)P_m(x)Z^{n+m} = (1-2xz+z^2)^{-1}$

Integrating both sides with respect to x from -1 to 1 , we get

$\sum_{n=0}^{\infty} \sum_{m=0}^{\infty} (\int_{-1}^{1} P_n(x)P_m(x)dx)Z^{n+m} = \int_{-1}^{1} (1-2xz+z^2)^{-1}dx$ ....(7)

Now from (4) and (7)

$\sum_{n=0}^{\infty} \int_{-1}^{1} ((P_n(x))^2 dx)Z^{2n} = \int_{-1}^{1} \frac{dx}{1-2xz+z^2} = [\frac{log(1-2xz+z^2)}{-2z}]_{-1}^{1} = -\frac{1}{2z}[log(1-z)^2 - log(1+z)^2]$

$= \frac{1}{z}[log(1+z) - log(1-z)] = \frac{2}{z}(z + \frac{z^3}{3} + \frac{z^5}{5}) = \frac{2}{z}\sum_{n=0}^{\infty} \frac{z^{2n+1}}{2n+1}$

Equating coefficient z^{2n} ,we get $\sum_{n=0}^{\infty} \int_{-1}^{1} ((P_n(x))^2 dx) = \frac{2}{2n+1}$ ....(8)

Combining (4) and (8) , we get $\int_{-1}^{1} P_m(x)P_n(x)dx = \begin{cases} 0 & \text{if } m \neq n \\ \\ \frac{2}{2n+1} & \text{if } m = n \end{cases}$

Q-17 Prove that $nP_n = (2n-1)xP_{n-1} - (n-1)P_{n-2}; n \geq 2$

We know that generating function $(1-2xz+z^2)^{-\frac{1}{2}} = \sum_{n=0}^{\infty} Z^n P_n(x)$ ....

Differentiating both sides with respect to z of (1)

$-\frac{1}{2}(1-2xz+z^2)^{-\frac{3}{2}}(-2x+2z) = \sum_{n=0}^{\infty} nZ^{n-1}P_n(x)$ ....(2)

$\Rightarrow (1-2xz+z^2)^{-\frac{3}{2}}(x-z) = \sum_{n=0}^{\infty} nZ^{n-1}P_n(x)$

Multiplying both sides by $(1-2xz+z^2)$

$(x-z)(1-2xz+z^2)^{-\frac{1}{2}} = (1-2xz+z^2)\sum_{n=0}^{\infty} nZ^{n-1}P_n(x)$, Now from (1)

$(x-z)\sum_{n=0}^{\infty} Z^n P_n(x) = (1-2xz+z^2)\sum_{n=0}^{\infty} nZ^{n-1}P_n(x)$

$\Rightarrow x\sum_{n=0}^{\infty} Z^n P_n(x) - z\sum_{n=0}^{\infty} Z^n P_n(x)$

$= \sum_{n=0}^{\infty} n Z^{n-1} P_n(x) - 2xz \sum_{n=0}^{\infty} n Z^{n-1} P_n(x) + z^2 \sum_{n=0}^{\infty} n Z^{n-1} P_n(x)$

$\Rightarrow x \sum_{n=0}^{\infty} Z^n P_n(x) - \sum_{n=0}^{\infty} Z^{n+1} P_n(x)$

$= \sum_{n=0}^{\infty} n Z^{n-1} P_n(x) - 2x \sum_{n=0}^{\infty} n Z^n P_n(x) + \sum_{n=0}^{\infty} n Z^{n+1} P_n(x)$

$\Rightarrow x \sum_{n=0}^{\infty} Z^n P_n(x) - \sum_{n=0}^{\infty} Z^n P_{n-1}(x)$

$= \sum_{n=0}^{\infty} (n+1) Z^n P_{n+1}(x) - 2x \sum_{n=0}^{\infty} n Z^n P_n(x) + \sum_{n=0}^{\infty} (n-1) Z^{n+1} P_{n-1}(x)$

Equating coefficient of z^n on both sides , we get

$x P_n(x) - P_{n-1}(x) = (n+1) P_{n+1}(x) - 2xn P_n(x) + (n-1) P_{n-1}(x)$

$(n+1) P_{n+1}(x) = (2n+1) x P_n(x) - n P_{n-1}(x)$ ....(3)

Replacing n by $n-1$ in (3)

$n P_n = (2n-1) x P_{n-1} - (n-1) P_{n-2}$

Q-18 Prove that $J_n(x+y) = \sum_{k=-\infty}^{\infty} J_{n-k}(x) J_k(y)$ and hence deduce that $1 + J_0^2(x) + 2 \sum_{k=1}^{\infty} J_k^2(x)$

We know that Generating function for the Bessel's function

$exp \frac{1}{2} x(z - \frac{1}{z}) = \sum_{n=-\infty}^{\infty} J_n(x)$ ....(1)

From (1) $exp \frac{1}{2} (x+y)(z - \frac{1}{z}) = \sum_{n=-\infty}^{\infty} J_n(x+y) z^n$ ....(2)

Now , $exp \frac{1}{2} x(z - \frac{1}{z}) exp \frac{1}{2} y(z - \frac{1}{z}) = \sum_{r=-\infty}^{\infty} \sum_{k=-\infty}^{\infty} J_r(x) J_k(y) z^{r+k}$

For a fixed value of r , we get z^n by taking $r + k = n$ i.e. $k = n - r$, $r = n - k$

Thus,keeping k fixed the coefficient of z^n in (1) is $J_{n-k}(x) J_k(y)$

So, the total coefficient of z^n will be given by summing all such term

from $k = -\infty$ to $k = \infty$ and it is given by $\sum_{k=-\infty}^{\infty} J_{n-k}(x) J_k(y)$

$\therefore exp[\frac{1}{2}(x+y)(z - \frac{1}{z})] = exp[\frac{1}{2} x(z - \frac{1}{z})] exp[\frac{1}{2} y(z - \frac{1}{z})] = \sum_{k=-\infty}^{\infty} J_{n-k}(x) J_k(y) z^n$

$\therefore \sum_{n=-\infty}^{\infty} z^n J_n(x) = \sum_{k=-\infty}^{\infty} J_{n-k}(x) J_k(y) z^n$ ....(3)

Now , equating coefficient of z^n from both sides of (3)

$J_n(x+y) = \sum_{k=-\infty}^{\infty} J_{n-k}(x) J_k(y)$

We have $exp[\frac{1}{2} x(z - z^{-1})] exp[\frac{1}{2} x(z - z^{-1})] = 1$

Which can be written as $\sum_{k=-\infty}^{\infty} J_k(x) z^k \sum_{k=-\infty}^{\infty} J_k(-x) z^k = 1$

$\Rightarrow [....... + J_2(x) t^{-2} + J_1(x) t^{-1} + J_0(x) + J_1(x) t + J_2(x) t^2 + ...][....... + J_2(x) t^{-2} - J_1(x) t^{-1} +$

$J_0(x) - J_1(x) t + J_2(x) t^2 - ....] = 1$

On comparing the constant terms on both sides, we get

$1 + J_0^2(x) + 2 \sum_{k=1}^{\infty} J_k^2(x)$

Q-19 Show that all the roots of $P_n(x)$ are real and lies between -1 and 1.

Let $f(x) = (x^2 - 1)^n = (x-1)^n(x+1)^n$ ....(1)

$\Rightarrow f(x)$ vanishes for $x = 1$ and $x = -1$ Hence by Roll's theorem $f'(x)$ must vanish at least once for some value α of x lying between -1 and 1. Now from (1),

we have $f'(x) = n(x-1)^{n-1}(x+1)^n + n(x-1)^n(x+1)^{n-1}$ ....(2)

(2) shows that $f'(x)$ vanishes at $x = 1$ and $x = -1$. But we have just proved that $f'(x)$ vanishes at $x = \alpha$ where $-1 < \alpha < 1$

Hence applying Roll's theorem to function f' two times. we conclude that $f''(x)$ must vanish at $x = \beta$ such that $-1 < \beta < \alpha$ and also at $x = \gamma$ such that $\alpha < \gamma < 1$ proceeding likewise we conclude that $f^{(n)}(x) = 0$ must have n real roots lying between -1 and 1 Using (1) , Rodreigue's formula gives

$P_n(x) = \frac{1}{2^n n!} \frac{d^n}{dx^n}(x^2 - 1)^n = \frac{1}{2^n n!} f^{(n)}(x)$ ....(3)

But $f^{(n)}(x) = 0 \Rightarrow P_n(x) = 0$

Since $f^{(n)}(x) = 0$ has a real roots between -1 and 1

Q-20 Prove that $\frac{d}{dx} F(\alpha, \beta; \gamma; x) = \frac{\alpha\beta}{\gamma} F(\alpha + 1, \beta + 1; \gamma + 1; x)$.

we know that definition of hyper geometric function

$F(\alpha, \beta; \gamma; x) = \sum_{r=0}^{\infty} \frac{(\alpha)_r (\beta)_r}{(\gamma)_r} \frac{x^r}{r!}$

Differentiating both sides with respect to x

$\frac{d}{dx} F(\alpha, \beta; \gamma; x) = \sum_{r=0}^{\infty} \frac{(\alpha)_r (\beta)_r}{(\gamma)_r} \frac{r x^{r-1}}{r!} = \sum_{r=1}^{\infty} \frac{(\alpha)_r (\beta)_r}{(\gamma)_r} \frac{x^{r-1}}{(r-1)!}$ $(\because r = 0$ term vanishes)

Take $r = m + 1 \Rightarrow m = r - 1$ if $r = 1 \Rightarrow m = 0$ and $r = \infty \Rightarrow m = \infty$

$\therefore \frac{d}{dx} F(\alpha, \beta; \gamma; x) = \sum_{m=0}^{\infty} \frac{(\alpha)_{m+1}(\beta)_{m+1}}{(\gamma)_{m+1}} \frac{x^m}{m!} = \sum_{m=0}^{\infty} \frac{\alpha(\alpha+1)_{m+1}\beta(\beta+1)_{m+1}}{\gamma(\gamma+1)_{m+1}} \frac{x^m}{m!}$

$\therefore \frac{d}{dx} F(\alpha, \beta; \gamma; x) = \frac{\alpha\beta}{\gamma} \sum_{m=0}^{\infty} \frac{(\alpha+1)_{m+1}(\beta+1)_{m+1}}{(\gamma+1)_{m+1}} \frac{x^m}{m!}$

$\therefore \frac{d}{dx} F(\alpha, \beta; \gamma; x) = \frac{\alpha\beta}{\gamma} F(\alpha + 1, \beta + 1; \gamma + 1; x)$

Q-21 Prove that $\int x^3 J_0(x) dx = x^3 J_1(x) - 2x^2 J_2(x) + c$

We know that $\frac{d}{dx} x^n J_n(x) = x^n J_{n-1}(x) \Rightarrow \int x^n J_{n-1}(x) = x^n J_n(x)$ ....(1)

Now $\int x^3 J_0(x) = \int x^2(x J_0(x)) dx = x^2 \int x J_0(x) dx - \int 2x(x J_1(x)) dx + c$

$\therefore \int x^3 J_0(x) = \int x^2(x J_0(x)) dx = x^3 \int J_0(x) dx - \int 2x(x J_1(x)) dx + c$

$\therefore \int x^3 J_0(x) = x^2(x J_1(x)) - \int 2x(x J_1(x)) dx + c$

$\therefore \int x^3 J_0(x) = x^2(x J_1(x)) - 2 \int (x^2 J_1(x)) dx + c$

$$\therefore \int x^3 J_0(x) = x^3 J_1(x) - 2x^2 J_2(x)dx + c$$

Q-22 Prove that $J_m(x) = \frac{1}{\pi}\int_0^\pi cos(m\theta - xsin\theta)d\theta$ and hence deduce that $J_0(x) = \frac{2}{\pi}\int_0^{\frac{\pi}{2}} cos(xsin\theta)d\theta$

We have results, $\int_0^\pi cosm\theta cosn\theta d\theta = \int_0^\pi sinm\theta sinn\theta d\theta = \begin{cases} 0 & \text{if } m \neq n \\ \frac{\pi}{2} & \text{if } m = n \end{cases}$ ....(1)

$$cos(xsin\theta) = J_0 + 2J_2 cos2\theta + 2J_4 cos4\theta + \dots \dots ....(2)$$

$$sin(xsin\theta) = 2J_1 sin\theta + 2J_3 sin3\theta + 2J_5 sin5\theta + \dots \dots ....(3)$$

Multiplying bothsides of (3) by $sinn\theta$ and then taking integrating with respect to θ between

limits 0 to π and using (1) we get, $\int_0^\pi sin(xsin\theta)sinn\theta d\theta = \begin{cases} 0 & \text{if n is even} \\ \pi J_n & \text{if n is odd} \end{cases}$ ....(4)

Again multiplying bothsides of (2) by $cosn\theta$ and then taking integration with respect to θ

between limits 0 to π and using (1),

we get $\int_0^\pi cos(xsin\theta)cosn\theta d\theta = \begin{cases} 0 & \text{if n is odd} \\ \pi J_n & \text{if n is even} \end{cases}$ ....(5)

Adding (4) and (5) for n is odd then $\int_0^\pi [cos(xsin\theta)cosn\theta + sin(xsin\theta)sinn\theta]d\theta = \pi J_n$

$\Rightarrow \int_0^\pi [cos(xsin\theta)cosn\theta]d\theta = \pi J_n \Rightarrow \int_0^\pi [cos(xsin\theta)cosn\theta]d\theta = \pi J_n(x)$

$n = m$

$\int_0^\pi [cos(xsin\theta)cosm\theta]d\theta = \pi J_m(x) \Rightarrow J_m(x) = \frac{1}{\pi}\int_0^\pi [cos(xsin\theta)cosm\theta]d\theta$

Integrating (2) with respect to θ between the limits 0 to π and using the result

$\int_0^\pi [cos(p\theta)d\theta = 0$; if p is even integer

We have $\int_0^\pi [cos(xsin\theta)d\theta = J_0(x)\int_0^\pi d\theta + 0 + 0 + \dots \dots = J_0(x)\pi$

$\therefore J_0(x) = \frac{1}{\pi}[cos(xsin\theta)d\theta$

Q-23 Find the Fourier Legendre expansion of a function defined by $f(x) = \begin{cases} 0; & \text{if } -1 << x < 0 \\ xJ_n; & \text{if } 0 << x << 1 \end{cases}$

We know that Fourier Legendre expansion of a function $f(x)$ is

$f(x) = \sum_{r=0}^\infty C_r P_r(x)$; where

$C_r = (r+\frac{1}{2})\int_{-1}^1 f(x)dx = \frac{2r+1}{2}[\int_0^1 f(x)P_r(x)dx + \int_{-1}^0 f(x)P_r(x)dx] = \frac{2r+1}{2}\int_0^1 xP_r(x)dx$..............(1)

Putting $r = 0, 1, 2, 3, \dots \dots$ in (1) , we get

$C_0 = \frac{1}{2}\int_0^1 xP_0(x)dx = \frac{1}{2}\int_0^1 xdx = \frac{1}{2}[\frac{x^2}{2}]_0^1 = \frac{1}{4}$

$C_1 = \frac{3}{2}\int_0^1 xP_1(x)dx = \frac{3}{2}\int_0^1 x^2 dx = \frac{3}{2}[\frac{x^3}{3}]_0^1 = \frac{1}{2}$

$C_2 = \frac{5}{2}\int_0^1 xP_2(x)dx = \frac{5}{2}\int_0^1 \frac{3x^3-x}{2}dx = \frac{5}{4}[\frac{3x^4}{4} - \frac{x^2}{2}]_0^1 = \frac{5}{8}$

$C_3 = \frac{7}{2}\int_0^1 xP_3(x)dx = \frac{7}{2}\int_0^1 x\frac{5x^2-2x}{2}dx = \frac{7}{4}\int_0^1 [5x^3 - 2x^2]dx = \frac{7}{4}[\frac{5x^4}{4} - \frac{2x^3}{3}]_0^1 = \frac{7}{4}[\frac{5}{4} - \frac{2}{3}] = \frac{49}{48}$

$\vdots$

$\vdots$

$\vdots$

Using these values in (1) , we get

$f(x) = \frac{1}{4}P_0(x) + \frac{1}{2}P_1(x) + \frac{5}{8}P_2(x) + \frac{49}{48}P_3(x) + \ldots\ldots\ldots$

Q-24 Find the third approximation of the solution of the equation $\frac{dy}{dx} = z$, $\frac{dz}{dx} = x^2z + x^4y$ by Picard's method, $y = 5$ and $z = 1$ when $x = 0$.

For $\frac{dy}{dx} = f(x, y, z) = z$, $\frac{dz}{dx} = g(x, y, z) = x^2z + x^4y$

If $y = y_0, z = z_0$ when $x = x_0$ then the successive approximation (y_n, z_n) is given by,

$y_n = y_0 + \int_{x_0}^x f(x, y_{n-1}, z_{n-1})dx$ ....(1)

$z_n = z_0 + \int_{x_0}^x g(x, y_{n-1}, z_{n-1})dx$ ....(2)

Here $f(x, y, z) = z, g(x, y, z) = x^2z + x^4y, x_0 = 0, y_0 = 5, z_0 = 1$

$y_n = y_0 + \int_0^x z_{n-1}dx$ ....(3)

$z_n = z_0 + \int_0^x (x^2z_{n-1} + x^4y_{n-1})dx$ ....(4)

For first approximation putting $n = 1$ in (3) , we get $y_1 = y_0 + \int_0^x z_0dx = 5 + \int_0^x 1dx = 5 + x$

$z_1 = z_0 + \int_0^x (x^2z_0 + x^4y_0)dx = 1 + \int_0^x (x^2 + 5x^4)dx = 1 + \frac{x^3}{3} + x^5$

For second approximation putting $n = 2$ in (3) , we get

$y_2 = y_0 + \int_0^x z_1dx = 5 + \int_0^x (1 + \frac{x^3}{3} + x^5)dx = 5 + x + \frac{x^4}{12} + \frac{x^6}{6}$

$z_2 = z_0 + \int_0^x [(x^2(1 + \frac{x^3}{3} + x^5) + x^4(5 + x)]dx = 1 + \int_0^x (x^2 + \frac{x^5}{3} + x^7 + 5x^4 + x^5)dx$

$\therefore z_2 = 1 + \frac{x^3}{3} + \frac{x^6}{3*6} + \frac{x^8}{8} + \frac{5x^5}{5} + \frac{x^6}{6} = 1 + \frac{2}{9}x^6 + \frac{x^3}{3} + x^5 + \frac{x^8}{8}$

For third approximation putting $n = 3$ in (3) , we get

$y_3 = y_0 + \int_0^x z_2dx = 5 + \int_0^x (1 + \frac{2}{9}x^6 + \frac{x^3}{3} + x^5 + \frac{x^8}{8})dx = 5 + x + \frac{x^4}{12} + \frac{x^6}{6} + \frac{2x^7}{63} + \frac{x^9}{72}$

$z_3 = z_0 + \int_0^x [(x^2(1 + \frac{2}{9}x^6 + \frac{x^3}{3} + x^5 + \frac{x^8}{8}) + x^4(5 + x + \frac{x^4}{12} + \frac{x^6}{6})]dx = 1 + \int_0^x (x^2 + \frac{x^5}{3} + x^7 + 5x^4 + x^5)dx$

$\therefore z_3 = 1 + \frac{x^{11}}{88} + \frac{x^{11}}{66} + \frac{x^9}{108} + \frac{x^8}{8} + \frac{x^6}{18} + x^5 + \frac{x^3}{3}$

Q-25 Solve by Picard's method

(I) $y' = y; y(0) = 1$

$\frac{dy}{dx} = y; y_0 = 1, x_0 = 0$ ....(1) then the successive approximation y_n is given by,

$y_n = y_0 + \int_{x_0}^{x} f(x, y_{n-1})dx$ ....(2)

$y_n = 1 + \int_{0}^{x} y_{n-1}dx$ ....(3)

For first approximation putting $n = 1$ in (3) , we get $y_1 = y_0 + \int_{0}^{x} y_0 dx = 1 + \int_{0}^{x} 1dx = 1 + x$

For second approximation putting $n = 2$ in (3) , we get

$y_2 = y_0 + \int_{0}^{x} y_1 dx = 1 + \int_{0}^{x}(1+x)dx = 1 + x + \frac{x^2}{2}$

For third approximation putting $n = 3$ in (3) , we get

$y_3 = y_0 + \int_{0}^{x} y_2 dx = 1 + \int_{0}^{x}(1 + x + \frac{x^2}{2})dx = 1 + x + \frac{x^2}{2} + \frac{x^3}{2*3}$

For fourth approximation putting $n = 4$ in (3) , we get

$y_4 = y_0 + \int_{0}^{x} y_3 dx = 1 + \int_{0}^{x}(1 + x + \frac{x^2}{2} + \frac{x^3}{6})dx = 1 + x + \frac{x^2}{2!} + \frac{x^3}{3!} + \frac{x^4}{4!}$

(II) $y' = x + y; y(0) = 1$

$\frac{dy}{dx} = x+y; y_0 = 1, x_0 = 0$ ....(1) then the successive approximation y_n is given by,

$y_n = y_0 + \int_{x_0}^{x} f(x, y_{n-1})dx$ ....(2)

$y_n = 1 + \int_{0}^{x}(x + y_{n-1})dx$ ....(3)

For first approximation putting $n = 1$ in (3) , we get

$y_1 = y_0 + \int_{0}^{x}(x + y_0)dx = 1 + \int_{0}^{x}(x + 1)dx = 1 + x + \frac{x^2}{2}$

For second approximation putting $n = 2$ in (3) , we get

$y_2 = y_0 + \int_{0}^{x}(x + y_1)dx = 1 + \int_{0}^{x}(x + 1 + x + \frac{x^2}{2})dx = 1 + x + x^2 + \frac{x^3}{2*3}$

For third approximation putting $n = 3$ in (3) , we get

$y_3 = y_0 + \int_{0}^{x}(x + y_2)dx = 1 + \int_{0}^{x}(1 + x + x^2 + \frac{x^3}{2*3})dx = 1 + x + \frac{x^2}{2} + \frac{x^3}{2*3} + \frac{x^4}{2*3*4}$

Q-26 Find three successive approximations of the solution of $\frac{dy}{dx} = e^x + y^2, y(0) = 0$

$\frac{dy}{dx} = e^x+y^2; y_0 = 0, x_0 = 0$ ....(1) then the successive approximation y_n is given by,

$y_n = y_0 + \int_{x_0}^{x} f(x, y_{n-1})dx$ ....(2)

$y_n = 0 + \int_{0}^{x}(e^x + y_{n-1}^2)dx$ ....(3)

For first approximation putting $n = 1$ in (3) , we get

$y_1 = \int_0^x (e^x + y_0^2)dx = \int_0^x (e^x)dx = e^x - 1$

For second approximation putting $n = 2$ in (3) , we get

$y_2 = y_0 + \int_0^x (e^x + (e^x - 1)^2)dx = \int_0^x (e^x + e^{2x} - 2e^x + 1)dx = [e^x + \frac{e^{2x}}{2} - 2e^x + x]_0^1 = \frac{1}{2}[e^x - 2e^x + 2x + 1]$

For third approximation putting $n = 3$ in (3) , we get

$y_3 = y_0 + \int_0^x (e^x + y_2^2)dx = \int_0^x (e^x + \frac{1}{2}[e^x - 2e^x + 2x + 1]^2)dx$

$\therefore y_3 = \frac{1}{4}\int_0^x [e^4 x - 4e^3 x + 2e^2 x + 4x^2 + 4x + 1 + 4xe^2 x - 8xe^x]dx$

Q-27 Determine which of the following equations are integrable and find the solution of those which are integrable

(I) $(y^2 + xz)dx + (x^2 + yz)dy + 3z^2 dz = 0$

Comparing with $Pdx + Qdy + Rdz = 0$ ....(1) $P = y^2 + xz, Q = x^2 + yz, R = 3z^2$

We have integrability condition $P(\frac{\partial Q}{\partial z} - \frac{\partial R}{\partial y}) + Q(\frac{\partial R}{\partial x} - \frac{\partial P}{\partial z}) + R(\frac{\partial P}{\partial y} - \frac{\partial Q}{\partial x}) = 0$ ....(2)

$(y^2 + xz)y + (x^2 + yz)(-x) + 3z^2(2y - x) = y^3 - x^3 - 6yz^2 - 6xz^2 \neq 0$

So, equation is not integrable

(II) $y(1 + z^2)dx - x(1 + z^2)dy + (x^2 + y^2)dz = 0$

Comparing with $Pdx + Qdy + Rdz = 0$ ....(1) $P = y(1 + z^2), Q = -x(1 + z^2), R = x^2 + y^2$

We have integrability condition $P(\frac{\partial Q}{\partial z} - \frac{\partial R}{\partial y}) + Q(\frac{\partial R}{\partial x} - \frac{\partial P}{\partial z}) + R(\frac{\partial P}{\partial y} - \frac{\partial Q}{\partial x}) = 0$ ....(2)

$(y(1 + z^2))(-2xz - 2y) + (-x(1 + z^2))(2x - 2yz) + (x^2 + y^2)(1 + z^2 + 1 + z^2) = 0$

Showing that the condition of integrability is satisfied and hence the given equation is integrable

The auxiliary equation is $\frac{dx}{y(1+z^2)} = \frac{dx}{-x(1+z^2)} = \frac{dx}{x^2+y^2}$

From first two fractions $-xdx = ydy \Rightarrow x^2 + y^2 = c^2 = u$

From third two fractions $\frac{dz}{c^2} \Rightarrow dz = c^2 \Rightarrow z = 2c = v$

$Adu + Bdv = 0 \Rightarrow Ad(x^2 + y^2) + Bdz = 0 \Rightarrow 2xAx + 2yAdy + Bdz = 0$

Comparing with (1) $2xA = y(1 + z^2), 2yA = x(1 + z^2), B = x^2 + y^2$

$\Rightarrow A = \frac{y}{2x}(1 + z^2), B = c^2 = u$

$vdu + udv = 0 \Rightarrow v = cu \Rightarrow z = c(x^2 + y^2)$

Q-28 Prove that $\frac{d}{dx}[x^{-p}J_p(x)] = -x^{-p}J_{p+1}(x); n \geq 1$ and hence show that

(I) $J_{p+1}(x) = \frac{2p}{x}J_p(x) - J_{p-1}(x)$

We know that $J_p(x) = \sum_{r=0}^{\infty}(-1)^r \frac{1}{r!\Gamma(p+r+1)}\left(\frac{x}{2}\right)^{2r+p}$

Multiplying both sides by x^{-p} and then taking derivative with respect to x

$\frac{d}{dx}[x^{-p}J_p(x)] = x^{-p}\frac{d}{dx}\left[\sum_{r=0}^{\infty}(-1)^r \frac{1}{r!\Gamma(p+r+1)}\left(\frac{x}{2}\right)^{2r+p}\right]$

$\Rightarrow \frac{d}{dx}[x^{-p}J_p(x)] = \sum_{r=0}^{\infty}(-1)^r \frac{1}{r!\Gamma(p+r+1)}\left(\frac{1}{2}\right)^{2r+p}\frac{d}{dx}[x^{2r}]$

$\Rightarrow \frac{d}{dx}[x^{-p}J_p(x)] = \sum_{r=0}^{\infty}\frac{(-1)^r}{r(r-1)!}\frac{2rx^{2r-1}}{\Gamma(p+r+1)}\frac{1}{2^{2r+p}}$

$\Rightarrow \frac{d}{dx}[x^{-p}J_p(x)] = \sum_{r=1}^{\infty}\frac{(-1)^r}{(r-1)!}\frac{x^{2r-1}}{\Gamma(p+r+1)}\frac{1}{2^{2r+p-1}}$ $(\because r = 0$ term vanishes$)$

$Take\ r = m+1 \Rightarrow m = r - 1$; $m = 0 \Rightarrow r = 1$ and $m = \infty \Rightarrow r = \infty$

$\Rightarrow \frac{d}{dx}[x^{-p}J_p(x)] = -x^{-p}\sum_{m=0}^{\infty}\frac{(-1)^m}{m!\Gamma(p+m+2)}\left(\frac{x}{2}\right)^{2m+p+1}$

Take $m = r \Rightarrow \frac{d}{dx}[x^{-p}J_p(x)] = -x^{-p}\sum_{m=0}^{\infty}\frac{(-1)^r}{r!\Gamma(p+1+r+1)}\left(\frac{x}{2}\right)^{2r+p+1}$

So, $\frac{d}{dx}[x^{-p}J_p(x)] = -x^{-p}J_{p+1}(x)$

(II) $J_p'(x) = \frac{1}{2}[J_{p-1}(x) - J_{p+1}(x)]$

From reccurence relation

$J_p'(x) = J_{p-1}(x) - \frac{p}{x}J_p(x)$ ....(1)

$J_p'(x) = \frac{p}{x}J_p(x) - J_{p+1}(x)$ ....(2)

Adding (1) and (2) , we get $2J_p'(x) = [J_{p-1}(x) - J_{p+1}(x)]$

$\therefore J_p'(x) = \frac{1}{2}[J_{p-1}(x) - J_{p+1}(x)]$

Q-29 Prove that

(I) $\frac{d}{dx}F(\alpha, \beta; \gamma; x) = \frac{\alpha\beta}{\gamma}F(\alpha + 1, \beta + 1; \gamma + 1; x)$.

we know that definition of hyper geometric function

$F(\alpha, \beta; \gamma; x) = \sum_{r=0}^{\infty}\frac{(\alpha)_r(\beta)_r}{(\gamma)_r}\frac{x^r}{r!}$ ....(1)

Differentiating both sides with respect to x

$\frac{d}{dx}F(\alpha, \beta; \gamma; x) = \sum_{r=0}^{\infty}\frac{(\alpha)_r(\beta)_r}{(\gamma)_r}\frac{rx^{r-1}}{r!} = \sum_{r=1}^{\infty}\frac{(\alpha)_r(\beta)_r}{(\gamma)_r}\frac{x^{r-1}}{(r-1)!}$ $(\because r = 0$ term vanishes$)$

Take $r = m + 1 \Rightarrow m = r - 1$ if $r = 1 \Rightarrow m = 0$ and $r = \infty \Rightarrow m = \infty$

$\therefore \frac{d}{dx}F(\alpha, \beta; \gamma; x) = \sum_{m=0}^{\infty}\frac{(\alpha)_{m+1}(\beta)_{m+1}}{(\gamma)_{m+1}}\frac{x^m}{m!} = \sum_{m=0}^{\infty}\frac{\alpha(\alpha+1)_{m+1}\beta(\beta+1)_{m+1}}{\gamma(\gamma+1)_{m+1}}\frac{x^m}{m!}$

$\therefore \frac{d}{dx}F(\alpha, \beta; \gamma; x) = \frac{\alpha\beta}{\gamma}\sum_{m=0}^{\infty}\frac{(\alpha+1)_{m+1}(\beta+1)_{m+1}}{(\gamma+1)_{m+1}}\frac{x^m}{m!}$

$$\therefore \frac{d}{dx}F(\alpha, \beta; \gamma; x) = \frac{\alpha\beta}{\gamma}F(\alpha+1, \beta+1; \gamma+1; x)$$

(II) $F(\alpha, \beta; \gamma; x) = F(\beta, \alpha; \gamma; x)$

we know that definition of hyper geometric function

$$F(\alpha, \beta; \gamma; x) = \sum_{r=0}^{\infty} \frac{(\alpha)_r (\beta)_r}{(\gamma)_r} \frac{x^r}{r!}$$

From definition of hyper geometric function

$$F(\alpha, \beta; \gamma; x) = \sum_{r=0}^{\infty} \frac{((\beta)_r \alpha)_r}{(\gamma)_r} \frac{x^r}{r!} = F(\beta, \alpha; \gamma; x)$$

$$F(\alpha, \beta; \gamma; x) = F(\beta, \alpha; \gamma; x)$$

(III) $F(\alpha, \beta; \gamma; 0) = 0$

we know that definition of hyper geometric function

$$F(\alpha, \beta; \gamma; x) = \sum_{r=0}^{\infty} \frac{(\alpha)_r (\beta)_r}{(\gamma)_r} \frac{x^r}{r!}$$

From definition of hyper geometric function

$$F(\alpha, \beta; \gamma; 0) = \sum_{r=0}^{\infty} \frac{(\alpha)_r (\beta)_r}{(\gamma)_r} \frac{0^r}{r!} = 0$$

Q-30 Prove that

(I) $(1+x)^p = F(-p, b, b, x)$.

we know that definition of hyper geometric function

$$F(\alpha, \beta; \gamma; x) = \sum_{r=0}^{\infty} \frac{(\alpha)_r (\beta)_r}{(\gamma)_r} \frac{x^r}{r!} = 1 + \frac{(\alpha)(\beta)}{(\gamma)} \frac{x}{1!} + \frac{\alpha(\alpha+1)\beta(\beta+1)}{\gamma(\gamma+1)} \frac{x^2}{2!} + \dots$$

So, $F(-p, b; b; x) = 1 - px + \frac{-p(1-p)}{2!}x^2 + \dots = (1+x)^p$ ($\because$ Bionomial theorem)

(II) $e^x = \lim_{b \to \infty} F(a, b, a, \frac{x}{b})$

$$F(\alpha, \beta; \gamma; x) = \sum_{r=0}^{\infty} \frac{(\alpha)_r (\beta)_r}{(\gamma)_r} \frac{x^r}{r!} = 1 + \frac{(\alpha)(\beta)}{(\gamma)} \frac{x}{1!} + \frac{\alpha(\alpha+1)\beta(\beta+1)}{\gamma(\gamma+1)} \frac{x^2}{2!} + \dots$$

$$\Rightarrow F(a, b; a; x) = \sum_{r=0}^{\infty} \frac{(a)_r (b)_r}{(a)_r} \frac{x^r}{b^r r!} = \sum_{r=0}^{\infty} \frac{(b)_r}{b^r} \frac{x^r}{r!} = 1 + b\frac{x}{1!} + \frac{b(b+1)}{b^2} \frac{x^2}{2!} + \dots$$

Now applying $\lim_{b \to \infty}$ So , $\lim_{b \to \infty} F(a, b, a, \frac{x}{b}) = 1 + \frac{x}{1!} + \frac{x^2}{2!} + \frac{x^3}{3!} + \dots = e^x$

(III) $cos x = \lim_{a \to \infty} F(a, a, \frac{1}{2}, \frac{-x^2}{4a^2})$

$$F(\alpha, \beta; \gamma; x) = \sum_{r=0}^{\infty} \frac{(\alpha)_r (\beta)_r}{(\gamma)_r} \frac{x^r}{r!} = 1 + \frac{(\alpha)(\beta)}{(\gamma)} \frac{x}{1!} + \frac{\alpha(\alpha+1)\beta(\beta+1)}{\gamma(\gamma+1)} \frac{x^2}{2!} + \dots$$

$$\Rightarrow F(a, a; \frac{1}{2}; \frac{-x^2}{4a^2}) = \sum_{r=0}^{\infty} \frac{(a)_r (a)_r}{(\frac{1}{2})_r} \frac{(-\frac{x}{4a^2})^r}{r!} = 1 - \frac{x^2}{2!} + \frac{(1+\frac{1}{a})^2}{2*3*4}x^4 + \dots$$

Now applying $\lim_{a \to \infty}$

$$\lim_{a \to \infty} F(a, a, \frac{1}{2}, \frac{-x^2}{4a^2}) = 1 - \frac{x^2}{2!} + \frac{x^4}{4!} + \dots cos x$$

Q-31 Solve $p^2 x + q^2 = z$ by using Jacobi's method.

$$p^2 x + q^2 = z \dots (1)$$

Let solution of (1) be of the form $u(x, y, z) = 0$ ....(2)

Partially with respect to x and y respectively of (2) gives

$\frac{\partial u}{\partial x} + \frac{\partial u}{\partial z} * \frac{\partial z}{\partial x} = 0 \Rightarrow p_1 + p_3 * p = 0 \Rightarrow p = -\frac{p_1}{p_3}$ ....(3)

$\frac{\partial u}{\partial y} + \frac{\partial u}{\partial z} * \frac{\partial z}{\partial y} = 0 \Rightarrow p_2 + p_3 * q = 0 \Rightarrow q = -\frac{p_2}{p_3}$ ....(4)

where $p = \frac{\partial z}{\partial x}, q = \frac{\partial z}{\partial y}, p_1 = \frac{\partial u}{\partial x} = \frac{\partial u}{\partial x_1}, p_2 = \frac{\partial u}{\partial y} = \frac{\partial u}{\partial x_2}, p_3 = \frac{\partial u}{\partial z} = \frac{\partial u}{\partial x_3}$

Taking $x = x_1, y = x_2, z = x_3$ and using (3)

$(1) \Rightarrow (-\frac{p_1}{p_3})^2 x_1 + (-\frac{p_2}{p_3})^2 = x_3 \Rightarrow x_1 p_1^2 + p_2^2 - x_3 p_3^2 = 0$

Let $f(x_1, x_2, x_3, p_1, p_2, p_3) = (-\frac{p_1}{p_3})^2 x_1 + (-\frac{p_2}{p_3})^2 = x_3$

$\Rightarrow x_1 p_1^2 + p_2^2 - x_3 p_3^2 = 0$ ....(5)

Now Jacobi's auxiliary equations are

$\frac{dp_1}{\frac{\partial f}{\partial x_1}} = \frac{dx_1}{-\frac{\partial f}{\partial p_1}} = \frac{dp_2}{\frac{\partial f}{\partial x_2}} = \frac{dx_2}{-\frac{\partial f}{\partial p_2}} = \frac{dp_3}{\frac{\partial f}{\partial x_3}} = \frac{dx_3}{-\frac{\partial f}{\partial p_3}}$

$\Rightarrow \frac{dp_1}{p_1^2} = \frac{dx_1}{-2p_1 x_1} = \frac{dp_2}{0} = \frac{dx_2}{-2p_2} = \frac{dp_3}{-p_3^2} = \frac{dx_3}{-2p_3 x_3}$

Taking the first two fractions $\frac{2}{p_1} dp_1 + \frac{1}{x_1} dx_1 = 0$

Integrating $2 log p_1 + log x_1 = log a_1 \Rightarrow x_1 p_1^2 = a_1 \Rightarrow p_1 = (\frac{a_1}{x_1})^{\frac{1}{2}}$

Taking third and fourth fractions gives,

$p_2 dp_2 + \frac{1}{2} dx_2 = 0 \Rightarrow \frac{p_2^2}{2} + \frac{1}{2} x_2 = a_2 \Rightarrow p_2 = (2a_2 - x_2)^{\frac{1}{2}}$

Substituting values of p_1 and p_2 in (5)

$p_3^2 = \frac{1}{x_3}[x_1(\frac{a_1}{x_1}) + 2a_2 - x_2] \Rightarrow p_3 = [\frac{a_1 + 2a_2 - x_2}{x_3}]^{\frac{1}{2}}$

Putting the values of p_1, p_2, p_3 in $du = p_1 dx_1 + p_2 dx_2 + p_3 dx_3$

$\Rightarrow du = a_1^{\frac{1}{2}} x_1^{-\frac{1}{2}} dx_1 + (2a_2 - x_2)^{\frac{1}{2}} dx_2 + (a_1 + 2a_2 - x_2)^{\frac{1}{2}} x_3^{-\frac{1}{2}} dx_3$

Integrating, $u = 2(a_1 x)^{\frac{1}{2}} +$

Taking $a_2 = 1$ and using (4) , the required solution $u = 0$ is given by,

$u =$

Q-32 Solve $2(z + xp + yq) = yp^2$ by using Charpit's method.

Given equation is $f(x, y, z, p, q) = 2(z + px + qy) - yp^2 = 0$ ....(1)

Its Charpit's auxiliary equations are

$$\frac{dp}{f_x+pf_z} = \frac{dq}{f_y+qf_z} = \frac{dz}{-pf_p-qf_q} = \frac{dx}{-f_p} = \frac{dy}{-f_q} \Rightarrow \frac{dp}{2p+2p} = \frac{dq}{2q-p^2+2q} = \frac{dz}{-p(2x-2yp)-q2y} = \frac{dx}{-(2x-2yp)} = \frac{dy}{-2y}$$

Taking first and last fractions $\frac{dp}{4p} = \frac{dy}{-2y} \Rightarrow \frac{dp}{p} + 2\frac{dy}{-2y} = 0$

Integrating , $logp + 2logy = loga \Rightarrow py^2 = a$ ....(2)

Solving (1) and (2) for p and q

$p = \frac{a}{y^2}$ and $q = -\frac{z}{y} - \frac{ax}{y^3} + \frac{a^2}{2y^4}$

We know that $dz = pdx + qdy$

$\therefore dz = \frac{a}{y^2}dx + [-\frac{z}{y} - \frac{ax}{y^3} + \frac{a^2}{2y^4}]dy$

Multiplying bothsides by y and re - arranging, we get

$(ydz + zdy) - a(\frac{ydx-xdy}{y^2}) - \frac{a^2}{2y^3}dy = 0 \Rightarrow d(yz) - ad(\frac{x}{y}) - \frac{a^2}{2}y^{-3}dy = 0$

Integrating , $yz - a(\frac{x}{y}) + \frac{a^2}{4y^2} = b$; whrere a and b being arbitrary constants

Q-33 State and prove necessary and sufficient condition for compatibility of partial differential equation $f(x,y,z,p,q) = 0$ and $g(x,y,z,p,q) = 0$.

Consider first order partial differential equations

$f(x,y,z,p,q) = 0$ ....(1)

$g(x,y,z,p,q) = 0$ ....(2)

Both (1) and (2) are compatible when every solution of one is also a solution of the other.

To find condition for (1) and (2) to be compatible.

Let $J =$ Jacobian of f and $g = \frac{\partial(f,g)}{\partial(p,q)} \neq 0$ ....(3)

Then (1) and (2) can be solved to obtain the explicit expression for p and q given by

$p = \phi(x,y,z)$ and $q = \psi(x,y,z)$ ....(4)

The condition that the pair of equations (1) and (2) should be compatible reduces then to the condition that the system of equations (4) should be completely integrable.

i.e. That the equation $dz = pdx + qdy$ or

$\phi dx + \psi dy - dz = 0$ ($\because$ by (4)) ....(5)

should be integrable.

Here (5) is integrable if $\phi(\frac{\partial \psi}{\partial z} - 0) + \psi(0 - \frac{\partial \phi}{\partial z}) + (-1)(\frac{\partial \phi}{\partial y} - \frac{\partial \psi}{\partial x}) = 0$

Which is equaivalent to $\frac{\partial \psi}{\partial x} + \phi(\frac{\partial \psi}{\partial z}) = \frac{\partial \phi}{\partial y} + \psi(\frac{\partial \phi}{\partial z})$ ....(6)

Substituting from equations (4) in (1) and differentiating with respect to x and z respectively

$$\frac{\partial f}{\partial x} + \frac{\partial f}{\partial p}\frac{\partial \phi}{\partial x} + \frac{\partial f}{\partial q}\frac{\partial \psi}{\partial x} = 0 \text{ ....(7)}$$

$$\frac{\partial f}{\partial z} + \frac{\partial f}{\partial p}\frac{\partial \phi}{\partial z} + \frac{\partial f}{\partial q}\frac{\partial \psi}{\partial z} = 0 \text{ ....(8)}$$

From (7) and (8)

$$\frac{\partial f}{\partial x} + \phi(\frac{\partial f}{\partial z}) + \frac{\partial f}{\partial p}(\frac{\partial \phi}{\partial x} + \phi\frac{\partial \phi}{\partial z}) + \frac{\partial f}{\partial q}(\frac{\partial \psi}{\partial x} + \phi\frac{\partial \psi}{\partial z}) = 0 \text{ ....(9)}$$

Similarly, from (2)

$$\frac{\partial g}{\partial x} + \phi(\frac{\partial g}{\partial z}) + \frac{\partial g}{\partial p}(\frac{\partial \phi}{\partial x} + \phi\frac{\partial \phi}{\partial z}) + \frac{\partial g}{\partial q}(\frac{\partial \psi}{\partial x} + \phi\frac{\partial \psi}{\partial z}) = 0 \text{ ....(10)}$$

Solving (9) and (10) , we get $\frac{\partial \psi}{\partial x} + \phi\frac{\partial \psi}{\partial z} = \frac{1}{J}[\frac{\partial(f,g)}{\partial(x,p)} + \phi\frac{\partial(f,g)}{\partial(z,p)}]$ ....(11)

Again substituting from equation(4) in (1) and differentiating with respect to y and z , and procceding them as above , we obtain

$$\frac{\partial \phi}{\partial y} + \psi\frac{\partial \phi}{\partial z} = -\frac{1}{J}[\frac{\partial(f,g)}{\partial(y,q)} + \phi\frac{\partial(f,g)}{\partial(z,q)}] \text{ ....(12)}$$

Substituting from equations (11) and (12) in (1) and replacing ϕ, ψ by p, q respectively , we obtain $\frac{1}{J}[\frac{\partial(f,g)}{\partial(x,p)} + p\frac{\partial(f,g)}{\partial(z,p)}] = -\frac{1}{J}[\frac{\partial(f,g)}{\partial(y,q)} + q\frac{\partial(f,g)}{\partial(z,q)}]$ or $[f,g] = 0$ ....(13)

where $[f,g] = \frac{\partial(f,g)}{\partial(x,p)} + p\frac{\partial(f,g)}{\partial(z,p)} + \frac{\partial(f,g)}{\partial(y,q)} + q\frac{\partial(f,g)}{\partial(z,q)}$

Q-34 Solve $yzp + zxq = xy$ by Langrange's method.

Given equation is $yzp + zxq = xy$ ....(1)

The Lagrange's subsidiary equation for (1) are

$$\frac{dx}{yz} = \frac{dy}{zx} = \frac{dz}{xy} \text{ ....(2)}$$

From the first two fractions, $\frac{dx}{yz} = \frac{dy}{zx} \Rightarrow xdx - ydy = 0 \Rightarrow x^2 - y^2 = c_1$ ....(3)

From (3) $x^2 = c_1 + y^2$ and $x^2 - z^2 = c_1 + c_2$

From first and last fraction of (2) $\frac{dx}{yz} = \frac{dz}{xy}$ ....(4)

From (3) and (4) the required general integral is $\phi(x^2 - y^2, x^2 - z^2) = 0$; where ϕ being an arbitrary function

Q-35 Solve

(I) $2(x^2 + x)y'' + (1 + 5x)y' + y = 0; x = 0$

Given equation is $2(x^2 + x)y'' + (1 + 5x)y' + y = 0$

Dividing by $2(x^2 + x)$ we have $y'' + \frac{1+5x}{2(x^2+x)}y' + \frac{1}{2(x^2+x)}y = 0$ ....(1)

Comparing (1) with standard equation $y'' + P(x)y' + Q(x)y = 0$

here $P(x) = \frac{1+5x}{2(x^2+x)}$ and $Q(x) = \frac{1}{2(x^2+x)}$

So that $xP(x) = \frac{1+5x}{2(x+1)}$ and $x^2Q(x) = \frac{1}{2(x+1)}$

Showing that both $P(x)$ and $Q(x)$ are analytic at $x = 0$ is a regular singular point.

To find series expansion , we take Let the solution of (1) be

$y = c_0 + c_1x + c_2x^2 + c_3x^3 + \dots \dots = \sum_{m=0}^{\infty} c_m x^{k+m}$; $c_0 \neq 0$...(2)

Differentiating (3) twice in succession with respect to 'x' , we get

$y' = \sum_{m=0}^{\infty} c_m(k+m)x^{k+m-1}$ and $y'' = \sum_{m=0}^{\infty} c_m(k+m)(k+m-1)x^{k+m-2}$ ....

Putting the above values of y, y' and y'' in (1)

$2(x^2 + x)\sum_{m=0}^{\infty}(k + m)(k + m - 1)c_m x^{k+m-2} + (1 + 5x)\sum_{m=0}^{\infty} c_m(k + m)x^{k+m-1} + \sum_{m=0}^{\infty} c_m x^m = 0$

$\Rightarrow 2\sum_{m=0}^{\infty}(k+m)(k+m-1)c_m x^{k+m} + 2\sum_{m=0}^{\infty} c_m(k+m)x^{k+m-1} + \sum_{m=0}^{\infty} c_m(k+m)x^{k+m-1}$

$\sum_{m=0}^{\infty} c_m x^{k+m} + 5\sum_{m=0}^{\infty} c_m(k + m)x^{k+m} + \sum_{m=0}^{\infty} c_m x^m = 0$

$\sum_{m=0}^{\infty}[(2(k + m)(k + m - 1)) + (k + m)]c_m x^{k+m-1} + \sum_{m=0}^{\infty} c_m[2(k + m)(k + m - 1) + 5(k + m) + 1]x^{k+m} = 0$...............(4)

Which is an identity equating to zero the coefficient of the smallest power in x, namely x^k gives the indicial equation.

$c_0 k(2k - 1) = 0$; as $c_0 \neq 0 \Rightarrow k = 0, \frac{1}{2}$...............(5)

Now we equate to zero coefficient of x^{k+m}

$c_m(k + m)[2(k + m - 1) + 1] + c_{m-1}[2(k + m - 1)(k + m - 2) + 5(k + m - 1) + 1] = 0$

$\Rightarrow c_m = -\frac{[(k+m-1)(2k+m-2)+5]+1}{(k+m)[2(k+m-1)+1]}c_{m-1}$ ....(6)

Taking $m = 1, 2, 3, \dots \dots \dots$ in (6), gives

$c_1 = -\frac{[k(2k-1)+5]+1}{(k+1)[2k+1]}c_0$

$c_2 = -\frac{[(k+1)2k+5]+1}{(k+2)[2(k+1)+1]}c_1 \Rightarrow c_2 = -\frac{[(k+1)2k+5]+1}{(k+2)[2(k+1)+1]}\frac{[k(2k-1)+5]+1}{(k+1)[2k+1]}c_0$.........................

Put these values in (2) $y = x^k[c_0 + c_1x + c_2x^2 + c_3x^3 + \dots \dots \dots]$

$\Rightarrow y = x^k[c_0 + [-\frac{[k(2k-1)+5]+1}{(k+1)[2k+1]}c_0]x + [-\frac{[(k+1)2k+5]+1}{(k+2)[2(k+1)+1]}\frac{[k(2k-1)+5]+1}{(k+1)[2k+1]} + \dots \dots \dots]$

$\Rightarrow y = c_0 x^k[1 + [-\frac{k}{(k+1)(2k+1)-1}]x + [-\frac{k(k+1)}{[(k+2)(2k+3)-1][(k+1)(2k+1)-1]}]x^2 c_0]x^2 + \dots \dots \dots]$...........................

Putting $k = 0$ and replacing c_0 by a in (7),

$y = ax^0[1 + [-\frac{5+1}{1}]x + \frac{(5+1)(5+1)}{(2)[(2)+1]}x^2 + \ldots\ldots\ldots]$

$\Rightarrow y = a[1 + 6x + \frac{(36)}{6}x^2 + \ldots\ldots\ldots]$

$\Rightarrow y = ax[1 - 6x + 6x^2 + \ldots\ldots\ldots] = au$

Putting $k = \frac{1}{2}$ and replacing c_0 by b in (7)

$y = bx^{\frac{1}{2}}[1 - \frac{11}{3}x + \frac{15 \cdot 11}{120}x^2 + \ldots\ldots\ldots] = bv$

So, the required general solution is $y = au + bv$

$y = ax[1 - 6x + 6x^2 + \ldots\ldots\ldots] + bx^{\frac{1}{2}}[1 - \frac{11}{3}x + \frac{15 \cdot 11}{120}x^2 + \ldots\ldots\ldots]$

(II) $(x^2 - x - 6)y'' + (5 + 3x)y' + y = 0; x = 3.$

Given equation is $(x^2 - x - 6)y'' + (5 + 3x)y' + y = 0$ ....(1)

Dividing by $(x^2 - x - 6)$ we have $y'' + \frac{5+3x}{(x^2-x-6)}y' + \frac{1}{(x^2-x-6)}y = 0$ ....(2)

Comparing (2) with standard equation $y'' + P(x)y' + Q(x)y = 0$

here $P(x) = \frac{5+3x}{(x^2-x-6)}$ and $Q(x) = \frac{1}{(x^2-x-6)}$

So that $xP(x) = \frac{5+3x}{(x-3)(x+2)}$ and $x^2Q(x) = \frac{1}{(x-3)(x+2)}$

So that $(x-3)P(x) = \frac{5+3x}{(x+2)}$ and $(x-3)^2Q(x) = \frac{1}{(x+2)}$

Showing that both $P(x)$ and $Q(x)$ are analytic at $x = 3$ is a regular singular point.

To find series expansion , we take Let the solution of (1) be

$y = c_0 + c_1(x-2)^k + c_2(x-2)^{k+2} + c_3(x-2)^{k+3} + \ldots\ldots\ldots$

$y = \sum_{m=0}^{\infty} c_m(x-2)^{k+m}$; $c_0 \neq 0$..........................(2)

Let $x - 3 = t$ then $\frac{dy}{dx} = \frac{dy}{dt}\frac{dt}{dx} = \frac{dy}{dt}$

$\Rightarrow \frac{d^2y}{dx^2} = \frac{d}{dx}(\frac{dy}{dx}) = \frac{d}{dt}(\frac{dy}{dx})\frac{dt}{dx} = \frac{d^2y}{dt^2}$ $\therefore$ series solution is $\sum_{m=0}^{\infty} c_m t^{k+m}$ ....(3)

$y' = \sum_{m=0}^{\infty} c_m(k+m)t^{k+m-1}$ and $y'' = \sum_{m=0}^{\infty} c_m(k+m)(k+m-1)t^{k+n-2}$ ....(4)

Putting the above values of y, y' and y'' in (1)

$((t-2)^2 - (t-2) - 6)y'' + (5 + 3(t-2))y' + y = 0$

$\sum_{m=0}^{\infty}[(k+m)(k+m-1)]c_m t^{k+m} - 5\sum_{m=0}^{\infty}(k+m)(k+m-1)c_m t^{k+m-1} - \sum_{m=0}^{\infty} c_m t^{k+m-1} +$

$\sum_{m=0}^{\infty} c_m t^{k+m} = 0$

$\sum_{m=0}^{\infty}[(k+m)(k+m-1) + 3(k+m) + 1]c_m t^{k+m} + \sum_{m=0}^{\infty} -5(k+m)(k+m-1) - (k+m)c_m t^{k+m-}$

0 ....(4)

Which is an identity equating to zero the coefficient of the smallest power in t, namely t^k gives the indicial equation.

$c_0 - 5k(k-1) - k = 0$; as $c_0 \neq 0 \Rightarrow k = 0, \frac{4}{5}$ ....(5)

Now we equate to zero coefficient of t^{k+m}

$c_m[(k+m)(k+m-1)+3(k+m)+1] + c_m + [5(k+m-1)(k+m) - (k+m+1)]C_{m+1} = 0$

$\Rightarrow c_{m+1} = -\frac{[(k+m)(k+m-1)+3(k+m)+1]}{[5(k+m-1)(k+m)-(k+m+1)]}c_m$ ....(6)

Taking $m = 1, 2, 3, ...$in (6), gives

$c_1 = -\frac{[k+1]}{5k+1}c_0$

$c_2 = -\frac{(k^2+4k+4)(k+1)}{(5k+6)(k+2)(5k+1)}c_0$

Put these values in (2) $y = c_0 t^k[c_0 + c_1 x + c_2 x^2 + c_3 x^3 + ...]$

$\Rightarrow y = t^k[c_0 + [-\frac{[k+1]}{5k+1}c_0]t + [\frac{(k^2+4k+4)(k+1)}{(5k+6)(k+2)(5k+1)}c_0]t^2 + ...$ ....(7)

Putting $k = 0$ and replacing c_0 by a in (7),

$\Rightarrow y = a[1 + t + \frac{1}{3}t^2] = a[1 + (x-2) + \frac{1}{3}(x-2)^2 + ...] = au$

Putting $k = \frac{4}{5}$ and replacing c_0 by b in (7)

$y = bt^{\frac{4}{5}}[1 + \frac{9}{25}t + \frac{196 \cdot 9}{625}t^2 + ...] = b(x-2)^{\frac{4}{5}}[1 + \frac{9}{25}(x-2) + \frac{196 \cdot 9}{625}(x-2)^2 + ...] = bv$

So, the required general solution is $y = au + bv$

$y = a[1 + (x-2) + \frac{1}{3}(x-2)^2 + ...] + b(x-2)^{\frac{4}{5}}[1 + \frac{9}{25}(x-2) + \frac{196 \cdot 9}{625}(x-2)^2 + ...]$

Q-36 Find complete integral of $q = 3p^2$

Given that $q = 3p^2$ ....(1)

Since (1) is of the form $f(p, q) = 0$, its Charpit's Auxiliary equations are

$\frac{dp}{f_x + pf_z} = \frac{dq}{f_y + qf_z} = \frac{dz}{-pf_p - qf_q} = \frac{dx}{-f_p} = \frac{dy}{-f_q}$

By (1) giving that $\frac{dp}{0} = \frac{dq}{0}$ taking the first ratio $dp = 0 \Rightarrow p = a = constant$

substituting in (1) , we get $f(a, q) = 0$ giving that $q = b = constant$

where $b = 3a^2$ such that $f(a, b) = 0$ then $dz = pdx + qdy$

Integrating , $z = ax + by + c$.........................(1); where$b$=3a^2$ and c is an arbitrary constant

From (1) , the complete integral is $z = ax + 3a^2y + c$

Q-37 Solve $z = px + qy + p^2 + q^2$ by using Charpit's method.

Let $f(x, y, z, p, q) = z - px - qy - p^2 - q^2 = 0$ ....(1)

Charpit's Auxiliary equations are

$$\frac{dp}{f_x+pf_z} = \frac{dq}{f_y+qf_z} = \frac{dz}{-pf_p-qf_q} = \frac{dx}{-f_p} = \frac{dy}{-f_q}$$ ....(2)

From (1) $f_x = -p, f_y = -q, f_z = 0, f_x = -x - 2p, f_q = -y - 2q$ ....(3)

Using (2) and (3)

$$\frac{dp}{0} = \frac{dq}{0} = \frac{dz}{p(x+2p)+q(y+2q)} = \frac{dx}{x+2p} = \frac{dy}{y+2q}$$ ....(4)

From the first fraction $dp = 0 \Rightarrow p = a$ and second fraction $dq = 0 \Rightarrow q = b$

putting $p = a, q = b$ in (1)

$z = ax + by + a^2 + b^2$ where a, b are being constant

Q-38 Find the general solution of

(a) $(D^2 - D'^2 - 3D + 3D')z = x^2 + 2y$

The Auxiliary Equation of given equation is given by

$m^2 - 1 - 3m + 3 = 0 \Rightarrow m^2 - 3m + 2 = 0 \Rightarrow (m - 2)(m - 1) = 0 \Rightarrow m = 2, 1$

$\therefore C.F. = \phi_1(y + x) + \phi_2(y + 2x)$; where ϕ_1, ϕ_2 being arbitrary functions

P.I.$=\frac{1}{D^2-D'^2-3D+3D'}(x^2 + 2y) = \frac{1}{D^2[1-\frac{D'^2}{D^2}-\frac{3}{D}+\frac{3D'}{D^2}]}(x^2 + 2y)$

$\therefore P.I. = \frac{1}{D^2}(1 - \frac{D'^2}{D^2} - \frac{3}{D} + \frac{3D'}{D^2})^{-1}(x^2 + 2y) = \frac{1}{D^2}(1 - (\frac{D'^2}{D^2} + \frac{3}{D} - \frac{3D'}{D^2}))^{-1}(x^2 + 2y)$

$\therefore P.I. = \frac{1}{D^2}(1+\frac{D'^2}{D^2}+\frac{3}{D}-\frac{3D'}{D^2}+........)(x^2+2y) = \frac{1}{D^2}(x^2+2y)+\frac{1}{D^4}(2)+\frac{1}{D^3}(3)-+\frac{1}{D^4}(3(2y))$

$\therefore P.I. = \frac{1}{D}\frac{x^3}{3} + 2xy + \frac{1}{D^3}(2x) + \frac{1}{D^2}(3x) - \frac{1}{D^3}(6xy) = \frac{x^4}{4} + x^2y + \frac{1}{D^2}x^2 + \frac{1}{D}\frac{3x^2}{2} - \frac{1}{D^2}\frac{6x^2y}{2}$

$\therefore P.I. = \frac{x^4}{4} + x^2y + \frac{x^4}{12} + \frac{x^3}{2} - \frac{x^4}{4}y$

Hence the required general solution is

$z = \phi_1(y + x) + \phi_2(y + 2x) + \frac{x^4}{3} - \frac{x^4}{4}y + \frac{x^3}{2} + x^2y$

(b) $DD'(2D - D' - 2)z = e^x(x - 2y)$

$\Rightarrow (2D^2D' - DD' - 2DD')z = e^x(x - 2y)$

The Auxiliary Equation of given equation is given by

$2m^2 - m - 2m = 0 \Rightarrow 2m^2 - 3m = 0 \Rightarrow m(2m - 3) = 0 \Rightarrow m = 0, \frac{3}{2}$

$\therefore C.F. = \phi_1(y)+\phi_2(y+\frac{3}{2}x) = \phi_1(y)+\phi_2(2y+3x)$; where ϕ_1, ϕ_2 being arbitrary functions

P.I.$=\frac{1}{DD'(2D-D'-2)}(x - 2y)e^x = \frac{1}{DD'(2D-D'-2)}e^x(x - 2y)$

$\therefore P.I. = \frac{1}{(2D^2-D'^2-2DD')}e^x(x - 2y)$

Hence the required general solution is

$$z = \phi_1(y) + \phi_2(y + \tfrac{3}{2}x) = \phi_1(y) + \phi_2(2y + 3x)$$

(c) $(D - D')(2D' - D + 1) = sin(x - y)$

$$2DD' - D^2 + D - 2D'^2 + DD' - D' = sin(x - y)$$

$$3DD' - D^2 + D - 2D'^2 - D' = sin(x - y)$$

The Auxiliary Equation of given equation is given by

$$3m - m^2 + m - 2 - 1 = 0 \Rightarrow m^2 - 4m + 3 = 0 \Rightarrow (m - 1)(m - 3) = 0 \Rightarrow m = 3, 1$$

$\therefore C.F. = \phi_1(y + x) + \phi_2(y + 3x)$; where ϕ_1, ϕ_2 being arbitrary functions

P.I.$= \frac{1}{(D-D')(2D'-D+1)} sin(x - y) = \frac{1}{(1+1)(2(-1)-1+1)} \int sinv\,du\,dv$; where $v = x - y$

$\therefore P.I. = \tfrac{1}{4} sinv = \tfrac{1}{4} sin(x - y)$

Hence the required general solution is

$$z = \phi_1(y + x) + \phi_2(y + 3x) + \tfrac{1}{4} sin(x - y)$$

(d) $(X^2D^2 - 4y^2D'^2 - 6yD')z = 0$

Let $x = e^u \Rightarrow u = logx, y = e^v \Rightarrow v = logy$ ....(1)

Let $D = \frac{\partial}{\partial x}, D' = \frac{\partial}{\partial y}, D_1 = \frac{\partial}{\partial u}, D_1' = \frac{\partial}{\partial v}$

Then the given equation becomes

$$(x^2D^2 - 4y^2D'^2 - 6yD')z = 0$$

$$\Rightarrow [D_1(D_1 - 1) - 4D_1'(D_1' - 1) - 6D_1']z = 0$$

$$\Rightarrow [D_1^2 - D_1 - 4D_1'^2 + 4D_1' - 6D_1']$$

$$\Rightarrow [D_1^2 - D_1 - 4D_1'^2 - 2D_1']z = 0$$

(e) $(D^2 + 2DD' - 8D')z = \sqrt{x + 2y}$

The Auxiliary Equation of given equation is given by

$$m^2 + 2m - 8 = 0 \Rightarrow (m + 4)(m - 2) = 0 \Rightarrow m = 2, -4$$

$\therefore C.F. = \phi_1(y + 2x) + \phi_2(y - 4x)$; where ϕ_1, ϕ_2 being arbitrary functions

P.I.$= \frac{1}{D^2 + 2DD' - 8D'^2} \sqrt{x + 2y}$

$\therefore P.I. = \frac{1}{1^2 + 2(1)(2) - 8(2)^2} \int \int \sqrt{v}\,dv\,dv$; where $v = x + 2y$

$\therefore P.I. = \frac{1}{1 + 4 - 32} \int \frac{v^{\frac{3}{2}}}{\frac{3}{2}} dv$

$\therefore P.I. = \frac{1}{27} \frac{2}{3} \int v^{\frac{3}{2}} dv$

$\therefore P.I. = \frac{4}{405}(x + 2y)^{\frac{5}{2}}$

Hence the required general solution is

$$z = \phi_1(y + 2x) + \phi_2(y - 4x) + \tfrac{4}{405}(x + 2y)^{\frac{5}{2}}$$

Q-39 Classify the equation and convert it into canonical form : $4r - y^6 t = 3y^5 q$

Given equation $4r - y^6 t = 3y^5 q$ ....(1)

Comparing (1) with $Rr + Ss + Tt + f(x, y, z, p, q) = 0$

Here $R = 4, S = 0, T = -y^6$

$\therefore S^2 - 4RT = 16y^6 > 0$, showing that (1) is hyperbolic

The λ - quadratic equation is :

$$R\lambda^2 + S\lambda + T = 0 \Rightarrow 4\lambda^2 - y^6 = 0 \Rightarrow \lambda = \tfrac{y^3}{2} \text{ and } \lambda = -\tfrac{y^3}{2}$$

The corresponding characteristic equation

$$\tfrac{dy}{dx} + \lambda_1 = 0 \Rightarrow \tfrac{dy}{dx} + \tfrac{y^3}{2} = 0 \dots\dots\dots\dots\dots\dots(2) \text{ and}$$

$$\tfrac{dy}{dx} + \lambda_1 = 0 \Rightarrow \tfrac{dy}{dx} - \tfrac{y^3}{2} = 0 \dots\dots\dots\dots\dots\dots(3)$$

Now , $\tfrac{dy}{dx} + \tfrac{y^3}{2} = 0 \Rightarrow \tfrac{dy}{y^3} = -\tfrac{dx}{2} \Rightarrow y^{-3} dy = -\tfrac{1}{2} dx$

Applying integration , $\Rightarrow \tfrac{y^{-2}}{-2} + x = c_1 \Rightarrow \tfrac{1}{y^2} + x = c_1$ and

$$\tfrac{dy}{dx} - \tfrac{y^3}{2} = 0 \Rightarrow \tfrac{dy}{y^3} = \tfrac{dx}{2} \Rightarrow y^{-3} dy = \tfrac{1}{2} dx$$

Applying integration , $\Rightarrow \tfrac{y^{-2}}{2} + x = c_1 \Rightarrow \tfrac{1}{y^2} - x = c_2$

Choose $u = x + \tfrac{1}{y^2}$ and $v = -x + \tfrac{1}{y^2}$

$$J = \tfrac{\partial(u,v)}{\partial(x,y)} = \begin{vmatrix} \tfrac{\partial u}{\partial x} & \tfrac{\partial u}{\partial x} \\ \tfrac{\partial v}{\partial x} & \tfrac{\partial v}{\partial y} \end{vmatrix} = \begin{vmatrix} 1 & -2y^{-3} \\ -1 & -2y^{-3} \end{vmatrix} = -4y^{-3} \neq 0$$

Now , $P = \tfrac{\partial z}{\partial x} = \tfrac{\partial z}{\partial u}\tfrac{\partial u}{\partial x} + \tfrac{\partial z}{\partial v}\tfrac{\partial v}{\partial x}$

$$\Rightarrow \tfrac{\partial z}{\partial x} = \tfrac{\partial z}{\partial u} - \tfrac{\partial z}{\partial v} \Rightarrow \tfrac{\partial}{\partial x} = \tfrac{\partial}{\partial u} - \tfrac{\partial}{\partial v}$$

$$q = \tfrac{\partial z}{\partial y} = \tfrac{\partial z}{\partial u}\tfrac{\partial u}{\partial y} + \tfrac{\partial z}{\partial v}\tfrac{\partial v}{\partial y} = -2y^{-3}\tfrac{\partial z}{\partial u} + (-2y^{-3})\tfrac{\partial z}{\partial v} = -2y^{-3}(\tfrac{\partial z}{\partial u} + \tfrac{\partial z}{\partial v})$$

$$r = \tfrac{\partial^2 z}{\partial x^2} = \tfrac{\partial}{\partial x}[\tfrac{\partial z}{\partial x}] = \tfrac{\partial}{\partial u}[\tfrac{\partial z}{\partial u}] - \tfrac{\partial}{\partial x}[\tfrac{\partial z}{\partial v}] = [\tfrac{\partial}{\partial u}[\tfrac{\partial z}{\partial u}]\tfrac{\partial u}{\partial x} + \tfrac{\partial}{\partial v}[\tfrac{\partial z}{\partial u}]\tfrac{\partial v}{\partial x}] - [\tfrac{\partial}{\partial u}[\tfrac{\partial z}{\partial v}]\tfrac{\partial u}{\partial x} + \tfrac{\partial}{\partial v}[\tfrac{\partial z}{\partial v}]\tfrac{\partial v}{\partial x}]$$

$$\therefore r = \tfrac{\partial^2 z}{\partial u^2} - \tfrac{\partial^2 z}{\partial v^2} - 2\tfrac{\partial^2 z}{\partial u \partial v} \dots\dots\dots\dots\dots\dots(4)$$

$$t = \tfrac{\partial^2 z}{\partial y^2} = \tfrac{\partial}{\partial y}[\tfrac{\partial z}{\partial y}] = \tfrac{\partial}{\partial y}(-2y^{-3})[\tfrac{\partial z}{\partial u} + \tfrac{\partial z}{\partial v}] = -2\tfrac{\partial}{\partial y}[y^{-3}\tfrac{\partial z}{\partial u}] + -2\tfrac{\partial}{\partial y}[y^{-3}\tfrac{\partial z}{\partial v}] + -2\tfrac{\partial}{\partial y}[y^{-3}\tfrac{\partial z}{\partial v}]$$

$$\therefore t = 6y^{-4}[\tfrac{\partial z}{\partial u} + \tfrac{\partial z}{\partial v}] + 4y^{-6}[\tfrac{\partial^2 z}{\partial u^2} + \tfrac{\partial^2 z}{\partial v^2}] + 8y^{-6}[\tfrac{\partial^2 z}{\partial u \partial v}]$$

$$\therefore 4[\tfrac{\partial^2 z}{\partial u^2} - \tfrac{\partial^2 z}{\partial v^2} - 2\tfrac{\partial^2 z}{\partial u \partial v}] - y^{-6}[6y^{-4}[\tfrac{\partial z}{\partial u} + \tfrac{\partial z}{\partial v}] + 4y^{-6}[\tfrac{\partial^2 z}{\partial u^2} + \tfrac{\partial^2 z}{\partial v^2}] + 8y^{-6}[\tfrac{\partial^2 z}{\partial u \partial v}]] = 3y^5[-2y^{-3}(\tfrac{\partial z}{\partial u} + \tfrac{\partial z}{\partial v})]$$

$$\Rightarrow \frac{\partial^2 z}{\partial v^2} + 2\frac{\partial^2 z}{\partial u \partial v} = 0$$

Q-40 Convert the equation into canonical form : $r + 2s + t = 0$

Given eqation $r + 2s + t = 0$ ....(1)

Comparing (1) with $Rr + Ss + Tt + f(x, y, z, p, q) = 0$

Here $R = 1, S = 2, T = 1$ so that $S^2 - 4RT = 4 - 4 = 0$, Showing that (1) is parabolic

The λ - quadratic equation is $R\lambda^2 + S\lambda + T = 0 \Rightarrow \lambda^2 + 2\lambda + 1 = 0 \Rightarrow \lambda = -1, -1$

The corresponding characteristic equation

$$\frac{dy}{dx} + \lambda_1 = 0 \Rightarrow \frac{dy}{dx} - 1 = 0 \Rightarrow \frac{dy}{dx} = 1 \Rightarrow dy = dx \Rightarrow x - y = c \text{ ....(2)} ;$$

where c being constant

Choose $u = x - y$ and $v = x + y$ where we have choosen $v = x + y$ in such a manner u and v are independent function as varified below

$$J = \frac{\partial(u,v)}{\partial(x,y)} = \begin{vmatrix} \frac{\partial u}{\partial x} & \frac{\partial u}{\partial x} \\ \frac{\partial v}{\partial x} & \frac{\partial v}{\partial y} \end{vmatrix} = 2 \neq 0$$

Now , $P = \frac{\partial z}{\partial x} = \frac{\partial z}{\partial u}\frac{\partial u}{\partial x} + \frac{\partial z}{\partial v}\frac{\partial v}{\partial x}$

$$\Rightarrow \frac{\partial z}{\partial x} = \frac{\partial z}{\partial u} - \frac{\partial z}{\partial v} \Rightarrow \frac{\partial}{\partial x} = \frac{\partial}{\partial u} - \frac{\partial}{\partial v}$$

$q = \frac{\partial z}{\partial y} = \frac{\partial z}{\partial u}\frac{\partial u}{\partial y} + \frac{\partial z}{\partial v}\frac{\partial v}{\partial y} = -\frac{\partial z}{\partial u} + \frac{\partial z}{\partial v}$

$r = \frac{\partial^2 z}{\partial x^2} = \frac{\partial}{\partial x}[\frac{\partial z}{\partial x}] = \frac{\partial}{\partial u}[\frac{\partial z}{\partial u}] - \frac{\partial}{\partial x}[\frac{\partial z}{\partial v}] = [\frac{\partial}{\partial u}[\frac{\partial z}{\partial u}]\frac{\partial u}{\partial x} + \frac{\partial}{\partial v}[\frac{\partial z}{\partial u}]\frac{\partial v}{\partial x}] - [\frac{\partial}{\partial u}[\frac{\partial z}{\partial v}]\frac{\partial u}{\partial x} + \frac{\partial}{\partial v}[\frac{\partial z}{\partial v}]\frac{\partial v}{\partial x}]$

$\therefore r = \frac{\partial^2 z}{\partial u^2} + \frac{\partial^2 z}{\partial v^2} + 2\frac{\partial^2 z}{\partial u \partial v}$ ....(4)

$t = \frac{\partial^2 z}{\partial y^2} = \frac{\partial}{\partial y}[\frac{\partial z}{\partial y}] = \frac{\partial}{\partial y}(-2y^{-3})[\frac{\partial z}{\partial u} + \frac{\partial z}{\partial v}] = -2\frac{\partial}{\partial y}[y^{-3}\frac{\partial z}{\partial u}] + -2\frac{\partial}{\partial y}[y^{-3}\frac{\partial z}{\partial v}] + -2\frac{\partial}{\partial y}[y^{-3}\frac{\partial z}{\partial v}]$

$\therefore t = \frac{\partial^2 z}{\partial u^2} + \frac{\partial^2 z}{\partial v^2} - 2\frac{\partial^2 z}{\partial u \partial v}$

$s = \frac{\partial^2 z}{\partial x \partial y} = \frac{\partial}{\partial x}[\frac{\partial z}{\partial y}] = [\frac{\partial}{\partial u} + \frac{\partial}{\partial v}][-\frac{\partial z}{\partial u} + \frac{\partial z}{\partial v}] = [\frac{\partial}{\partial u}][-\frac{\partial z}{\partial u} + [\frac{\partial z}{\partial v}]] + \frac{\partial}{\partial v}[-\frac{\partial z}{\partial u} + \frac{\partial z}{\partial v}] = -\frac{\partial^2 z}{\partial u^2} + \frac{\partial^2 z}{\partial v^2}$

$\therefore 4[\frac{\partial^2 z}{\partial u^2} - \frac{\partial^2 z}{\partial v^2} - 2\frac{\partial^2 z}{\partial u \partial v}] - y^{-6}[6y^{-4}[\frac{\partial z}{\partial u} + \frac{\partial z}{\partial v}] + 4y^{-6}[\frac{\partial^2 z}{\partial u^2} + \frac{\partial^2 z}{\partial v^2}] + 8y^{-6}[\frac{\partial^2 z}{\partial u \partial v}] = 3y^5[-2y^{-3}(\frac{\partial z}{\partial u} + \frac{\partial z}{\partial v})]$

$$\Rightarrow \frac{\partial^2 z}{\partial v^2} + 2\frac{\partial^2 z}{\partial u \partial v} = 0$$

Using value of r, s, t in (1) the required canonical form is $\frac{\partial^2 z}{\partial v^2} = 0$

Q-41 Using Monge's method , solve the equation : $r + s - 6t = 0$

Given equation is $r + s - 6t = 0$ ....(1)

Comparing it with $Rr + Ss + Tt = V$ we have $R = 1, S = 1, T = -6, V = 0$

Hence Monge's subsidary equations $Rdpdy + Tdqdx - Vdxdy = 0$

$\Rightarrow dpdy - 6dqdx = 0$ ....(2)

and $R(dy)^2 - sdxdy + T(dx)^2 = 0$

$(dy)^2 - dxdy - 6(dx)^2 = 0$ ....(3)

$dpdy - 6dqdx = 0$ and $(dy)^2 - dxdy - 6(dx)^2 = 0$

$\Rightarrow dy = [dx \pm (dx)^2 + 4 * 6(dx)^2{}^{\frac{1}{2}}]\frac{1}{2} = \frac{dx}{2}[1 \pm 5]$

$\Rightarrow dy = 3dx$ or $dy = -2dx$ ....(4) Integrating it $y - 3x = c_1; where c_2$ is constant

............................(5)

$y + 2x = c_2; where c_2$ is constant ....(6)

for $dy = 3dx$ from (1) $dp3dx - 6dqdx = 0 \Rightarrow dp = 2dq$

Integrating it $p - 2q = c_3$, where c_3 is constant ....(7)

for $dy = -2dx$ from (1) $-dp2dx - 6dqdx = 0 \Rightarrow dp + 3dq = 0$

Integrating it $p + 3q =_4; where c_4$ is constant ....(8)

From (5) and (7) ,

first intermediate integral is $p - 2q = f_1y - 3x$; f_1 is arbitrary function ....(9)

From (6) and (8) ,

first intermediate integral is $p + 3q = f_2y + 2x$; f_2 is arbitrary function ....(10)

Solving (9) and (10) for p and q

$p = \frac{1}{5}3f_1(y - 3x) + 2f_2(y + 2x)$ and $q = \frac{1}{6} - f_1(y - 3x) + f_2(y + 2x)$

Substituting these values of p and q in $dz = pdx + qdy$ we get,

$dz = \frac{1}{5}3f_1(y - 3x) + 2f_2(y + 2x)dx + \frac{1}{6} - f_1(y - 3x) + f_2(y + 2x)dy$

$\Rightarrow dz = f_1(y - 3x)d(\ frac35x - \frac{1}{6}y) + f_1(y - 3x)d(\frac{3}{5}x - \frac{1}{6}y)$

Integrating it , $z = F_1(y - 3x) + F_2(y + 2x)$; where F_1 and F_2 are arbitrary functions

Q-42 Using Monge's method , solve the equation : $rt - s^2 + 1 = 0$

Given equation is $(rt - s^2) = -1$ ....(1)

Comparing it with $Rr + Ss + Tt + U(rt - s^2) = V$ we have $R = 0, S = 0, T = 0, U = 1$ and $V = -1$ ....(2)

Here λ - quadratic equation $\lambda^2(UV + RT) + \lambda US + U^2 = 0$ ....(3)

becomes $\lambda^2 - 1 = 0 \Rightarrow \lambda = 1, -1 \Rightarrow \lambda_1 = -1, \lambda_2 = 1$ ....(4)

Since the two values of λ are distinct , we shall get two intermediate integrals , which are given by following sets of equations

$Udy + \lambda_1 Tdx + \lambda_1 Udp = 0$ and $Udx + \lambda_2 Rdy + \lambda_2 Udq = 0$ ....(5)

$Udy + \lambda_2 Tdx + \lambda Udp = 0$ and $Udx + \lambda_1 Rdy + \lambda_1 Udq = 0$ ....(6)

Using (2) and (4) , (5) and (6) reduces to $dy - dp = 0 \Rightarrow dp - dy = 0, dq + dx = 0$ ....(7)
$dp + dy = 0, dx - dq = 0 \Rightarrow dq - dx = 0$ ....(7)

Integrating of (6) and (7) respectively gives

$p - y = c_1, q + x = c_2, p + y = c_3$ and $q - x = c_4$ ....(8)

From (8) two intermidiate integral is given by , $p - y = f(q + x)$ and $p + y = F(q - x)$ ....(9)

Let $q + x = \alpha$ and $q - x = \beta$ ....(10)

From (9) $p - y = f(\alpha)$ and $p + y = F(\beta)$ ....(11)

In what follows α and β will be regarded as parameters solving (10) fot x and (11) for y , we have

$x = \frac{(\alpha - \beta)}{2}$ ....(12)

$y = \frac{(F(\beta) - f(\alpha))}{2}$ ....(13)

From (11) , $p = y + f(\alpha)$ ....(14)

From (10) , $q = x + \beta$ ....(15)

From (12) and (13) , $dx = \frac{1}{2}(d\alpha - d\beta)$ and $dy = \frac{1}{2}(d\alpha - d\beta)$

From (12) and (13) , $dx = \frac{1}{2}(d\alpha - d\beta)$ and $dy = \frac{1}{2}F'(\beta)d\beta - (f'(\alpha)d\alpha)$ ....(16)

$dz = pdx + qdy = [y + f(\alpha)]dx + [x + \beta]dy = [ydx + xdy] + f(\alpha)dx + \beta dy$

$\Rightarrow dz = d(xy) + f(\alpha)\frac{1}{2}(d\alpha - d\beta) + \beta\frac{1}{2}[F'(\beta)d\beta - f'(\alpha)d\alpha]$

$\Rightarrow dz = d(xy) + \frac{1}{2}f(\alpha)d\alpha - \frac{1}{2}[f(\alpha)d\beta + \beta f'(\alpha)d\alpha] + \beta F'(\beta)d\beta$

$\therefore 2dz = 2d(xy) + f(\alpha)d\alpha - [f(\alpha)d\beta + \beta f'(\alpha)d\alpha] + \beta F'(\beta)d\beta$ ....(17)

Let $\int f(\alpha)d\alpha = \phi(\alpha)$ and $\int f(\beta)d\beta = \psi(\beta)$ ....(18)

So that $f(\alpha) = \phi'(\alpha)$ and $F(\beta) = \psi'(\beta)$ ....(19)

Using (18) and (19) , (12) , (13) and (17) may be rewritten as

$2x = (\alpha - \beta)$,

$2y = \psi'(\beta) - \phi'(\alpha)$,

$2z = 2xy - \phi(\alpha) + \beta\phi'(\alpha) + \psi'(\beta) - \psi(\beta)$

Which is the required solution in parametric form , α and β being parameters and ϕ and ψ being arbitrary functions.

Q-43 By separating the variables , find the solution of three dimensional Laplace equation in cylindrical coordinate system.

Laplace's equation in cylindrical coordinates (ρ, ϕ, z) is

$\frac{\partial^2 u}{\partial \rho^2} + \frac{1}{\rho}\frac{\partial u}{\partial \rho} + \frac{1}{\rho^2}\frac{\partial^2 u}{\partial \phi^2} + \frac{\partial^2 u}{\partial z^2} = 0$ ....(1)

Let solution of (1) be of the form $u(\rho, \phi, z) = R(\rho)\Phi(\phi)Z(z)$ ....(2)

where R, ϕ and z are functions of ρ, ϕ and z respectively

Substituting this value of u in (1) , we get

$R''\phi z + \frac{1}{\rho}R'\phi z + \frac{1}{\rho}R\phi''z + R\phi z''$

Dividing throughout by $R\phi z$, we obtain

$\frac{R''}{R} + \frac{R'}{\rho R} + \frac{\phi''}{\rho^2\phi} = -\frac{z''}{z}$ ....(3)

Since the R.H.S. depends only on z while the L.H.S. depends only on ρ and ϕ , (3) can only true if each side is equal to the some constant say $-m^2$ then (3) gives

$z'' - m^2 z = 0$ ....(4) and

$$\frac{R''}{R} + \frac{R'}{\rho R} + \frac{\phi''}{\rho^2 \phi} = -m^2 \quad \text{............................(5)}$$

Re - writing (5) by separating variables , we have

$$\rho^2 \left(\frac{R''}{R} + \frac{R'}{\rho R} + m^2\right) = -\frac{\phi''}{\phi} \quad \text{............................(6)}$$

Since ρ and ϕ are independent , (6) is true only when each side is equal to the some constant on physical grounds of actual problems , we must involve trigonometric functions in solutions , so we chooses the proposed constant to be n^2 , furthermore in order to satisfy the physical condition

$$u(\rho, \phi) = u(\rho, \phi + 2\pi),$$

we supposed that n is an integer.

Then (6) reduces to $\phi'' + n^2 \phi = 0$ ....(7) and

$$\rho^2 R'' + \rho R' + (m^2 \rho^2 - n^2)R = 0$$

Solution of (4) , (7) and (8) are

$$Z(z) = Ae^{mz} + Be^{-mz}$$

$$\Phi(\phi) = C\cos n\phi + D\sin n\phi$$

$$R(\rho) = E_{mn}J_n(m\rho) + F_{mn}Y_n(mp)$$

Q-44 Solve wave equation in cartesian coordinates by method of separation variable and show that solution is $\psi(x, y, z, t) = e^{\pm i(lx + my + nz + kct)}$ where l, m, n, k are constant with $l^2 + m^2 + n^2 = k^2$.

Three dimensional wave equation is given by,

$$\frac{\partial^2 u}{\partial x^2} + \frac{\partial^2 u}{\partial y^2} + \frac{\partial^2 u}{\partial z^2} = \frac{1}{c^2}\left(\frac{\partial^2 u}{\partial t^2}\right) \quad \text{............................(1)}$$

Suppose (1) has solutions of the form $u(x, y, z, t) = X(x)Y(y)Z(z)T(t)$ ....(2)

Substituting this value of (2) in (1) and simplifying , we have

$$\frac{X''}{X} + \frac{Y''}{Y} + \frac{Z''}{Z} = \frac{1}{c^2}\frac{T''}{T} \quad \text{............................(3)}$$

Since x, y, z and t are independent variables (3) is true if each term on both sides is equal to a constant that is ,

$$\frac{X''}{X} = -l^2, \frac{Y''}{Y} = -m^2, \frac{Z''}{Z} = -n^2, \frac{1}{c^2}\frac{T''}{T} = -k^2 \quad \text{............................(4)}$$

Provided $k^2 = l^2 + m^2 + n^2$ ....(5)

Solutions of differential equations of (4) are of the forms

$$X(x) = e^{\pm ilx}, Y(y) = e^{\pm imy}, Z(z) = e^{\pm inz}, T(t) = e^{\pm ikt} \quad ....(6)$$

Using (2) and (6) , a solution of (1) can be put in the form

$$u(x, y, z, t) = e^{\pm i(lx+my+nz+kt)} \; ; \text{ where } k^2 = l^2 + m^2 + n^2$$

Q-45 If $(\beta D' + \gamma)^2$ is a factor of $F(D, D')$,then $e^{-\frac{\gamma}{\beta}y}[\phi_1(\beta X) + y\phi_2(\beta X)]$ is a solution of $F(D, D')z = 0$
, ϕ_1 and ϕ_2 are arbitrary functions of a single variable ξ.

Let $F(D, D')z = 0$ have a factor $(\beta D' + \gamma)^2$

Consider the equation $(\beta D' + \gamma)(\beta D' + \gamma)z = 0 \quad ....(1)$

Let $(\beta D' + \gamma)z = v \quad ....(2)$

then $(\beta D' + \gamma)v = 0 \quad ....(3)$

The general solution of (3) is $v = e^{-\frac{\gamma}{\beta}y}\phi(\beta x)$; where ϕ is a arbitrary function $.........................(4)$

Substituting from (4) is (2) , we have

$$(\beta D' + \gamma) = e^{-\frac{\gamma}{\beta}y}\phi(\beta x) \Rightarrow (\beta q + \gamma) = e^{-\frac{\gamma}{\beta}y}\phi(\beta x) \quad ....(5)$$

Lagrange's auxiliary equation of (5) are

$$\frac{dx}{0} = \frac{dy}{\beta} = \frac{dz}{e^{-\frac{\gamma}{\beta}y}\phi(\beta x) - \gamma z}$$

From first and second fractions

$$\beta dx = 0 dy \Rightarrow \beta x = \lambda$$

From first and third fractions

$$\frac{dx}{0} = \frac{dz}{e^{-\frac{\gamma}{\beta}y}\phi(\beta x) - \gamma z}$$

$$\Rightarrow \frac{dz}{dx} = e^{-\frac{\gamma}{\beta}y}\phi(\beta x) - \gamma z$$

$$\Rightarrow \frac{dz}{dx} + \gamma z = e^{-\frac{\gamma}{\beta}y}\phi(\lambda) \text{ it is linear differential equation.}$$

So, I.F. $= e^{\int \gamma dx} = e^{\gamma x}$

Solution is $ze^{\gamma x} = \int e^{-\frac{\gamma}{\beta}y}\phi(\lambda)e^{\gamma x}dx = e^{-\frac{\gamma}{\beta}y}\phi(\lambda)e^{\gamma x}\frac{1}{\gamma} + \mu$

$\therefore ze^{\gamma x} - e^{-\frac{\gamma}{\beta}y}\phi(\lambda)e^{\gamma x}\frac{1}{\gamma} = \mu$

$$\therefore ze^{\gamma x} - e^{\gamma x} e^{-\frac{\gamma}{\beta} y} \phi(\beta x) \frac{1}{\gamma} = \gamma \phi_1(\beta x)$$

$$z = e^{\gamma x} e^{-\frac{\gamma}{\beta} y} [\phi_1(\beta x) + y \phi_2(\beta x)]$$

References

1. "Higher Engineering Mathematics", B.S. Grewal, , Khanna Publishers, New Delhi, 2004.

2. "Advanced Differential Equations", M. D. Raisinghania, S. Chand & Co.

3. "Elementary Course in Partial Differential Equations", T. Amarnath, Narosa Publ. House, New Delhi.

4. "Special Functions", Saran, Sharma, Trivedi, Pragti Prakashan.

5. "Integral Calculus" , shanti Narayan , Dr. P.K.Mittal, S.Chand.

YOUR KNOWLEDGE HAS VALUE